THE GREEN
MARBLE

—

DAVID P. TURNER

THE GREEN MARBLE

Earth System Science

and Global Sustainability

COLUMBIA UNIVERSITY PRESS
NEW YORK

Columbia University Press
Publishers Since 1893
New York   Chichester, West Sussex
cup.columbia.edu
Copyright © 2018 Columbia University Press

Library of Congress Cataloging-in-Publication Data
Names: Turner, David (David P.), 1950– author.
Title: The green marble : earth system science and global sustainability /
David P. Turner.
Description: New York : Columbia University Press, 2018. |
Includes bibliographical references and index.
Identifiers: LCCN 2017039170 | ISBN 9780231180603 (cloth : alk. paper) |
ISBN 9780231180610 (pbk. : alk. paper) | ISBN 9780231542845 (e-book)
Subjects: LCSH: Global environmental change. | Biosphere. | Nature—Effect of
human beings on. | Environmental sciences—Social aspects. |
Environmental protection—International cooperation.
Classification: LCC GE149 .T87 2018 | DDC 304.2/8—dc23
LC record available at https://lccn.loc.gov/2017039170

Columbia University Press books are printed on permanent
and durable acid-free paper.
Printed in the United States of America

Cover design: Noah Arlow

CONTENTS

PREFACE

The title of this book, *The Green Marble*, refers to three perspectives on planet Earth.

The National Aeronautics and Space Administration (NASA) and the National Oceanic and Atmospheric Administration (NOAA) use the term *green marble* when referring to an image of Earth in which the mapped variable is an indicator of annual photosynthesis on land. Satellite remote sensing now provides information on a near daily basis for monitoring global photosynthesis. The green marble is a member of a pantheon of Earth marbles, including (1) the blue marble, a photo from space that emphasizes the preponderance of water on Earth's surface; (2) the black marble, Earth at night with its jewel-like network of lighted roads and cities; and (3) the white marble, snowball Earth from a period early in its geological history when it was largely covered by ice and snow. The emphasis—in the marble motif, and in this book—is on seeing Earth as a whole.

The second referent of the title is to the environmental movement. Green is often the adjective of choice to evoke environmentally friendly technologies, consumer choices, and natural resource management practices. Earth must inevitably become greener in that sense if the human enterprise is to prosper. Global environmental governance is a major theme in the later part of the book.

Lastly, and somewhat ominously, green sulfur bacteria were pervasive in the ocean during Earth's most extreme extinction event. About 250 million years ago, a massive pulse of greenhouse gases into the atmosphere (sound familiar?) drastically warmed the climate, changed the ocean circulation and chemistry, and induced a wave of species extinctions. Predominantly oxygen-loving red and blue-green bacteria were replaced by predominantly oxygen-intolerant green sulfur bacteria, an indicator of an environmental toxic to the majority of the species living at that time. This book examines the possibility of human instigation of another biosphere catastrophe.

The Green Marble is about the converging fields of Earth system science and global sustainability. After a chapter introducing the discipline of Earth system science, the next two chapters examine Earth's geological past. They describe the magnitude of variation in Earth's climate and biosphere over geological history and thus provide perspective on the current human perturbation of the Earth system. The next three chapters investigate the ongoing human disruption of Earth's biogeochemical cycles and its biosphere, as well as scenarios for Earth's environmental future. The last five chapters wrestle with what has been accomplished, and what can be done in the future, about global environmental governance.

The material here necessarily covers a lot of intellectual ground and freely crosses disciplinary boundaries. It is intended to be accessible to a non-geoscientist, but still achieve enough depth to give a real feel for the human predicament and the opportunities with respect to global environmental change. Citations to classic and contemporary literature are given to help unpack what may seem at times to be a flurry of abstractions. The goal is a synoptic coverage of global environmental change that will form a framework for organic growth of the reader's knowledge about global-scale structures and processes.

I feel immensely fortunate to be living in a time when electronic access to much of the scientific record is readily available. That heritage, along with ongoing interactions with colleagues and students, provided the foundation for this synthesis effort. I am especially grateful to colleagues for reviewing various sections of the book (Steve Frolking, David King, Denise Lach, David Rupp, and Mary Santelmann). During its long gestation, support was provided at times by the NASA Terrestrial Ecology Program, the NOAA Fellowship Program, and the Fulbright Scholar Program. I benefited significantly from the guidance of Patrick Fitzgerald, my editor at Columbia University Press. Special thanks to Mariette Brouwers, my fellow world traveler for 43 years.

Despite the kind reviews of my colleagues, I have undoubtedly made errors of fact, emphasis, and interpretation. These are certainly my responsibility and I appreciate the forbearance of all.

ABBREVIATIONS

AIS	automatic ship identification
AMOC	Atlantic meridional overturning circulation
AVHR	Advanced Very High Resolution Radiometer
BAU	business as usual
CCN	cloud condensation nuclei
CE	Common Era
CFC	chlorofluorocarbon
CPM	common pool resources
DGVM	dynamic global vegetation model
DMS	dimethyl sulfide
DNA	deoxyribonucleic acid
EM	ecological modernization
E/MSY	extinctions per million species years
EMT	ecological modernization theory
ENSO	El Niño–southern oscillation
EOS	NASA Earth Observing System
EPA	U.S. Environmental Protection Agency
EPME	end-Permian mass extinction
ESA	European Space Agency
ESM	Earth system model
FAO	United Nations Food and Agriculture Organization
FSC	Forest Stewardship Council
GCM	general circulation model
GDP	gross domestic product
GEF	Global Environmental Facility
GEOSS	Global Earth Observing System of Systems

GFW	Global Forest Watch
GPP	gross primary production
GRACE	Gravity Recovery and Climate Experiment
HCFC	hydrochlorofluorocarbon
IAM	integrated assessment model
ICESat	Ice, Cloud, and Land Elevation Satellite
IGOS	Integrated Global Observing System
IMF	International Monetary Fund
INGO	international nongovernmental organization
IPBES	Intergovernmental Science-Policy Platform on Biodiversity and Ecosystem Services
IPCC	Intergovernmental Panel on Climate Change
IUCN	International Union for the Conservation of Nature
IWC	International Whaling Commission
LEED	Leadership in Energy and Environmental Design
LUE	light use efficiency
MAT	mean annual temperature
MEA	millennium ecosystem assessment
MODIS	Moderate Resolution Imaging Spectroradiometer
NAFTA	North American Free Trade Agreement
NASA	U.S. National Aeronautics and Space Administration
NCAR	National Center for Atmospheric Research
NDVI	Normalized Difference Vegetation Index
NEE	net ecosystem exchange
NGO	nongovernmental organization
NMHC	non-methane hydrocarbons
NOAA	U.S. National Oceanic and Atmospheric Administration
NPP	net primary production
NSIDC	National Snow and Ice Data Center
PETM	Paleocene–Eocene thermal maximum
RCP	representative concentration pathway
REDD	Reduction in Deforestation and Forest Degradation
RNA	ribonucleic acid
SAL	structural adjustment lending
SES	socioecological system
THC	thermohaline circulation

UCB	UNCED Convention on Biodiversity
UNCCD	United Nations Convention to Combat Desertification
UNCED	United Nations Conference on the Environment and Development
UNEP	United Nations Environment Programme
UNFCCC	United Nations Framework Convention on Climate Change
UV	ultraviolet radiation
VIIRS	Visible Infrared Radiometer Suite
WEO	World Environmental Organization
WMO	World Meteorological Organization
WTO	World Trade Organization

THE GREEN MARBLE

1

EARTH SYSTEM SCIENCE

In every age there is a turning point, a new way of seeing and asserting the coherence of the world.

—J. Bronowski (1974)

THE CHALLENGE OF GLOBAL ENVIRONMENTAL CHANGE

Planet Earth can seem quite small. It takes only 90 minutes for an Earth-orbiting satellite to accomplish one global circuit. A single photograph can capture the entire Earth, with recognizable continents and oceans. Google Earth can take the virtual you nearly instantaneously to the top of Mount Everest or the depths of Death Valley.

Size is relative, of course, and in a sense our planet is rapidly becoming ever smaller. With the existing global transportation and communication networks, we can physically make our way to almost any location on the face of the planet within a day or two. The Internet allows us to communicate in text, sound, and images with billions of people around the world. The media keep us informed of breaking news locally and globally. In my professional life, satellite data give me near real-time information on weather and major biosphere disturbances everywhere on the planet.

The Earth is getting smaller in other fundamental ways. Until very recently, our species had only a minor influence on the global biogeo-chemical cycles of water, carbon, nitrogen, and other elements critical to life. However, in the past century we have become one of the dominant forces in those cycles (W. C. Clark, Crutzen, and Schellnhuber, 2004; Vitousek, Mooney, Lubchenco, and Melillo, 1997b). We are increasing the atmospheric carbon dioxide (CO_2) concentration by burning fossil fuels and converting forest land to cropland. Industrialized agriculture and fossil fuel combustion are now introducing into the environment more plant-available nitrogen per year than is produced by natural processes. We now use in one way or another 50 percent of global freshwater flow in streams and rivers. These anthropogenic impacts on the environment are changing the operation of the Earth system in ways unfavorable to advanced technological civilization (Barnosky et al., 2014). Perhaps more ominous in the long run than our influences on the global biogeochemical cycles is that we are driving species of plants and animals extinct at a vastly higher rate than is indicated in the paleorecord over much of evolutionary history.

As ecologists have said for decades, the problem starts with a human population that has been increasing for roughly the past 10,000 years—since the end of the last ice age. There are now over seven billion of us, and the projected balance of births and deaths will likely continue to favor population increase throughout this century. Equally significant, the ecological footprint per individual—i.e., per capita resource consumption—has increased dramatically in parallel with the number of people. About 20 percent of the global population has achieved a moderate to high standard of living (albeit leaving billions living quite precariously), but the cost in terms of environmental degradation has been enormous (MEA, 2005). If several billion additional people attain what is considered the modern lifestyle by exploiting resources in a manner similar to the lucky two billion, humanity will soon be living on a planet of weed-like species with a warmer climate than has been found on Earth in over 30 million years.

In recent decades, we have begun casting around for ways of thinking that might ameliorate our perverse impact on the environment and change the current trajectory. In broad terms, one aspect of the solution is to "think globally." That slogan traces back at least as far as the "think global, act local" motto of Friends of the Earth, an environmental organization founded by activist David Brower in 1969. "Think Global, Act

Local" was also the title of an influential essay published in 1972 by French microbiologist René Dubos (1901–1982). He promoted the concept at the United Nations Conference on the Human Environment in 1972.

The meaning of the "act local" part of the aphorism is straightforward. It means to work at saving specific geographic areas from environmental degradation. Friends of the Earth is built around the concept of people in specific areas organizing to protect those areas. To "think global" is the other side of the coin (Mol, 2000). It could be interpreted as an admonition to be aware of global-scale environmental problems, such as climate change, that are a manifestation of many local decisions. The expression serves as a reminder to foster solidarity on environmental issues among the wide variety of human cultures dispersed around the planet (N. Gough, 2002). However, to think at the global scale remains a challenge. In part this is likely because our brains were designed by biological evolution primarily to apprehend and respond to local events over short time frames, usually involving small groups of people and simple cause-and-effect relationships (Ehrlich and Ehrlich, 2009). Thinking about global environmental change inevitably requires us to stretch our capacity for imagination and for abstraction. This book is an effort to develop a foundation for thinking globally.

IMPACTS, FEEDBACKS, AND GOVERNANCE

We will pay special attention to three intertwined themes.

The first is human impacts. Earth has existed for more than four billion years. At the time of our recent arrival, it had evolved a well-established biosphere and a complex web of global biogeochemical cycles. However, humanity is rapidly changing how the Earth system operates. We are said to be entering a new geological period—the Anthropocene (Crutzen, 2002; Crutzen and Stoermer, 2000). In seeking to manage our global-scale impacts, we must understand the background functioning of the Earth system and the relative magnitude of our influences on it. Reciprocally, human-induced changes in the Earth system are beginning to negatively impact human welfare, and projections to 2100 and beyond suggest much worse to come.

Second is the concept of feedback. In common usage, the term refers to a reply or comment that conveys an evaluative message. A teacher gives students feedback on their essays. More formally, it refers to reciprocal interactions between different components of a system. In a negative feedback relationship, a change in one part of a system induces a change in another part that dampens the original change. This type of feedback is seen when an increase in the atmospheric CO_2 concentration induces an increase in biosphere photosynthesis, which then increases the rate of CO_2 uptake from the atmosphere and dampens the CO_2 rise. The alternative is a positive feedback relationship in which the original change induces a change in another part of the system that amplifies the original change. This is seen when climate warming causes more forest fires that increase emissions of greenhouse gases. With respect to the Earth system, we are especially interested in feedback relationships that regulate the global climate; these include biophysical processes that change with climate warming and serve to amplify or dampen the warming. The study of Earth's history and recent dynamics offers many clues.

The last theme is governance—specifically, global environmental governance. The meaning of "governance" is more inclusive than "government" because the former accounts for a broader range of actors in the self-regulation of a social body. A system of global environmental governance will ultimately include intergovernmental organizations, states, civil society, and engaged citizens. A critical question here is how to build an institutional framework to manage the relationship between humanity and the rest of the Earth system.

THINKING METAPHORICALLY

Considering our inherent difficulty in thinking globally, perhaps a good way to start is simply in terms of familiar similes and metaphors.

Probably the most deeply rooted metaphor about Earth is the vision of *Earth as mother*. We can intuit that cave painters of the prehistoric era, by going underground, were attempting to connect with the generative power of the Earth—possibly in the belief that through their drawings they were fertilizing it (Frankl, 2003) or releasing the animal spirits within (Lewis-Williams, 2002). We of course cannot know for sure what they were thinking, but even among extant hunter-gatherer cultures there is often a sentiment of reverence for Earth's fecundity. The mother metaphor is still used widely in contemporary culture: Mother Earth was referred to in the Paris Agreement on climate change, James Lovelock chose Gaia (the Greek goddess of Earth) as the name of his revolutionary hypothesis about planetary homeostasis (which we will examine in chapter 3), and one of historian Arnold Toynbee's mighty tomes was titled *Mankind and Mother Earth*.

Cultivation of plants for food began about 10,000 years ago, a step that introduced a fundamental change in the relationship of humans to their environment. In the hunter-gatherer era, the bounty of nature was there for the taking. With cultivation, the change was made to actively managing nature. At that point, the reigning metaphor became *Earth as garden*. This was not the Garden of Eden, but rather the garden that produces dinner. In a sense, cultivation began our separation from nature because we became the subject and Earth became the object that we sought to control. But cultivation also requires a certain intimacy with nature because we are motivated to understand it (albeit to exploit it more efficiently), and possibly protect it (to ensure a harvest).

The urge to understand nature, so as to better manipulate it, led to the discovery and rapid expansion of the scientific worldview in recent centuries. The dominant metaphor has thus become *Earth as machine*. Enlightenment-era French philosopher René Descartes (1596–1650) described both the human body and Earth itself as machines. In that view, we have a mostly instrumental relationship with the natural world and manipulate it as needed for our own objectives. The key feature of the machine metaphor is reductionism. We take nature apart, identify the mechanisms that drive it, and reorder them to meet our needs. A key weakness is that a machine is made by an agent outside the machine, but in the case of Earth, humanity is a part of the Earth system.

In recent decades, the consequences of the machine metaphor have become manifest on a global scale through pollution, natural resource degradation, and environmental change. The latest mythological figure evoked to characterize our relationship with the Earth is Medea (Ward, 2009). She was the mother in Greek mythology who killed her children.

Along with the Gaia hypothesis, we will consider the Medea hypothesis in chapter 3.

Alternative models that have emerged as antidotes to the reductionist thinking of the machine metaphor include *Earth as home* and *Earth as system*. The iconic image of sunlit Earth against a background of endless dark space (the "blue marble") has become a reminder of Earth's beauty and fragility. An especially poignant early photograph of Earth from space was made from the *Apollo 13* spacecraft as it limped back from an unsuccessful mission to the moon. Traditionally, a home is worth preserving and, if necessary, worth fighting for.

The power of the system metaphor is in identification of structures composed of parts and wholes, and finding the causal relationships and feedback relationships among the parts and between each whole and its environment. Humans are just one part of a larger system in this view. The emerging field of Earth system science aims to disentangle the hierarchy of parts and whole that make up the Earth system, to model the dynamics of this system's behavior, and to inform the evolution of a sustainable global civilization.

THE SEMANTICS OF THE SPHERES

One way to think about parts of the Earth system is in terms of spheres, and to continue our project of thinking globally in a more scientific sense, let's consider the sphere. Perhaps the most characteristic feature of Earth is its roughly spherical shape. This geometric form is quite common in nature and remarkable in any of its manifestations (Volk, 1995). The most startling thing about the sphere is its symmetry. From the perspective of energy balance and materials cycling, the sphere is a satisfying object of study because there is closure; i.e., it can readily be studied as a whole. An additional intriguing feature of a sphere is what happens on its surface as substances or energy forces grow and distribute themselves. First, density gradually increases; but once the surface is covered, the pressure of interaction increases, and the likelihood of new phenomena is enhanced.

Awareness in the Western world that we live on a sphere traces back to the Greek philosophers. Pythagoras (570–490 BCE) observed that mast tops appeared first when ships came into view on the horizon. Aristotle (384–322 BCE) knew that different stars became visible as a person traveled south, and observed that the Earth cast a curved shadow on the moon. Eratosthenes (276–194 BCE) estimated the circumference of Earth based on simple observations of shadows and use of trigonometry. Other early civilizations also had well-developed astronomical theories that included a spherical Earth. After the Copernican revolution in sixteenth-century western Europe (tracing back to Islamic astronomers), the idea that the Earth is a planet orbiting a star became part of the foundation of our global intellectual heritage.

The basics are that we live on a rock (the geosphere), the third rock from the sun. Earth's surface is two thirds water (the hydrosphere) and one third land, with 10 percent of the total land area currently covered by snow and ice (the cryosphere). Above us is the atmosphere, a layer of gases some 60 kilometers thick that consists mostly of an inert form of nitrogen, a reactive form of oxygen, and small concentrations of CO_2, methane, and other trace gases. Below us are the pedosphere (the soil) and the lithosphere (the Earth's crust). Several other spheres that greatly concern us in this book include the biosphere, the technosphere, and the noösphere. Later chapters explore these concepts in depth, but the following overview will briefly introduce them. A *Lexicon of the Spheres* appearing after chapter 11 defines these and other terms.

The Biosphere

Within the geophysical sciences tradition, the concept of the biosphere denotes the totality of life on Earth. The biosphere is an entity in the sense that it plays a major role in the global biogeochemical cycles. Life originated on the Earth relatively soon after the planet coalesced, between four and five billion years ago. Beginning about two billion years ago, there is evidence in the paleorecord of something that could be called a biosphere. By that time, microbial photosynthesis had led to a major change in the chemistry of the Earth's atmosphere—the conversion from an atmosphere most likely dominated by methane to one that included significant oxygen and CO_2. Since microorganisms were the source of the oxygen, this transformation was unequivocal evidence that living organisms were altering the physical and chemical environment at the global scale. As we shall see, the biosphere has strongly influenced the concentration of greenhouse gases in Earth's atmosphere throughout its history.

The spontaneous emergence of the biosphere from the geosphere was an extraordinary cosmic event (more about that in chapter 2). Such a thing could not be predicted from the laws of physics and chemistry as we know them (Kauffmann, 1995, 2008). James Lovelock noted the equally remarkable fact that the biosphere has gone on to survive for billions of years since its origin despite numerous devastating collisions with asteroids and comets; episodes of "Snowball Earth," when ice covered most or all of the planetary surface; episodes of much warmer climate than our current conditions; and a 25 percent increase in the intensity of solar radiation associated with an aging sun (Lovelock, 1979).

The biosphere is now enduring a form of disruption to which it has not previously been subjected. Notably, Earth's climate and the ocean chemistry are being rapidly altered, and extensive land cover change and resource exploitation are causing a wave of extinctions. The driving force this time is not the impact of an asteroid or an episode of intense volcanism, but instead the activities of a particularly industrious primate species known as *Homo sapiens*. Humans evolved like any other species, but developed a new trait—capacity for language—that initiated a new form of evolution, i.e., cultural evolution. Language introduced the ability to transmit information efficiently both vertically (across generations) and horizontally (among individuals across space), and that trait has created a new world order (McNeil, 2000). Whether you call it management or mismanagement, humans have begun to exert a strong influence on the Earth system as a whole.

The Technosphere

Although various animals use tools on occasion, modern humans have taken tool use to a level that is qualitatively different from anything else found in the animal kingdom. Over the last few hundred years, a layer of physical artifacts associated with tool use has gradually accumulated on Earth's surface (Zalasiewicz et al., 2016). This "technosphere" began with objects such as stone axes and primitive dwellings. However, as human populations grew and technology evolved, the technosphere has morphed into a ubiquitous mesh of buildings, machines, and infrastructure for communication and transportation. The image of the Earth at night (the "black marble") nicely captures the density of the technospheric web.

Again, from a geosciences perspective, we can think of the technosphere as a component of the Earth system, even as a kind of living thing. Cities are the organs of its body, highways are its blood vessels, and wires are the conduits of its nervous system. The technosphere ingests matter and energy. It consumes prodigious amounts of fossil fuels and appropriates a significant proportion of the biosphere's primary and secondary (i.e., consumer) production. Its outputs include the cornucopia of food and manufactured objects we associate with contemporary civilization. Its outputs also include large quantities of waste products, many of which are accumulating in the atmosphere, geosphere, and biosphere rather than being recycled (Haff, 2014). Of particular note, of course, is CO_2, a greenhouse gas and the product of fossil fuel combustion that is driving global climate change.

Like all living things, the technosphere grows. Although that growth was relatively slow from generation to generation over many centuries, it has greatly accelerated since World War II (Hibbard et al., 2007). This recent growth of the technosphere has raised standards of living for billions of people, but at a high cost to the biosphere. An increasingly urgent question confronts us: Can the technosphere be transformed into something that continues to provide goods and services to billions of people without degrading the biosphere and disrupting the climate system?

The Noösphere

French paleontologist and priest Pierre Teilhard de Chardin (1881–1955) is the writer most commonly associated with the development of the noösphere (*NOH-uh-sfeer*) concept (Sampson and Pitt, 1999; D. P. Turner, 2005). The "noos" part of the word is a Greek root referring to mind. For Teilhard, the noösphere was a sphere of mind, the totality of all human thought (Teilhard de Chardin, 1959/1955).

Teilhard's ideas about the noösphere were influenced by World War I (he was a stretcher-bearer). He saw that rapid advances in the technology of communication and transportation were in a sense "compressing" humanity. Teilhard observed that we live on the surface of a sphere and that the increasing density of humanity and the technology-driven intensification of interactions among us were creating a kind of psychic pressure. World War I was a manifestation of that pressure. Teilhard imagined that when the pressure became great enough, a transformation or unification would occur—the noösphere would coalesce.

Teilhard's book *The Phenomenon of Man* introduced the noösphere concept to a broad audience and was widely read. His writing received enthusiastic responses from some philosophers and scientists (e.g., Huxley, 1958), one appeal being its broad scope. Teilhard took a mostly scientific view of the history of the universe, and there is an undeniable philosophical comfort in his notion that humanity is a product of cosmological and biosphere evolution, thus in a sense at home in the universe. Systems theorist Stuart Kaufmann has elaborated on that perspective more recently (Kaufmann, 1995). In popular culture, Teilhard's noösphere concept has been evoked in relation to the development of the Internet (Kreisberg, 1995). His ideas about the tightening social integration across all of humanity, driven by advances in communication technology, fit well with the emerging role of the Internet as a reservoir of information and a means for global communication.

Russian biogeochemist Vladimir Vernadsky (1863–1945) conceived of the noösphere in a way that was radically different from Teilhard's concept. Vernadsky was one of the first scientists to quantitatively study the global biogeochemical cycles, and he offered a refreshing nondualistic interpretation of noösphere (Vernadsky, 1945). He recognized the growing magnitude and pervasiveness of human impacts on the planetary surface, likening them to a geological force. This new geological force was guided by mental phenomena rather than by strictly physical, chemical, or biological processes. Thus, a new way of characterizing the Earth system was needed. Vernadsky envisioned the noösphere as a new form of the biosphere, a biogeochemical cycling entity that included all life as well as its associated atmosphere and lithosphere. His noösphere was one dominated by human influences and serving primarily to meet the needs of humanity.

Vernadsky's noösphere concept, like Teilhard's, has received significant criticism. In the late 1950s, American ecologist Eugene Odum categorized the notion as "dangerous" because it implied that humanity was ready to take over management of the biosphere (Odum, 1959). Odum worked several decades after Vernadsky's death and was perhaps in a better position than Vernadsky to see that humanity's dominion over the biosphere might not be so benevolent. *Homo sapiens* needed no encouragement in believing it was wise enough to manage the Earth at a global scale. Anthropologist Gregory Bateson (1979) noted that there may well be a global system, but the nature of systems is such that a part (humanity) cannot control the whole (the Earth system). Nevertheless, the arrival of human consciousness does represent a qualitative change in how the Earth system functions, and it seems worthwhile to retain the term *noösphere* in that context. We may well wonder, too, if other noöspheres exist in the universe, in galaxies where self-aware life forms have evolved and come to manage (or mismanage as the case may be) their planetary biogeochemical cycles.

The Technobiosphere

Since Vernadsky's time, we have made considerable progress in quantifying the chemical pathways and the flux magnitudes of the global biogeochemical cycles. It is now obvious that human activities profoundly alter these cycles. In effect, there is a growing coupling of the technosphere and the biosphere. One might refer to this contemporary fusion as the "technobiosphere" (D. P. Turner, 2011). Energy inputs to the coupled system are a combination of solar energy and fossil fuels, and a key mode of interaction between the biosphere and the technosphere is the global carbon cycle. Anthropogenic (human-initiated) transfers of carbon to the atmosphere by way of fossil fuel combustion and deforestation have become a significant flux (i.e., flow) relative to background biologically driven fluxes such as global vegetation growth (Roy, Saugier, and Mooney, 2001). Much of the primary and secondary (consumer) productivity of the biosphere is now ingested by the technosphere in one way or another (Vitousek et al., 1997b). Through management of natural resources, the coupling of the biosphere and technosphere is becoming ever more integrated.

Whether the technobiosphere can become sustainable is an open question that we will examine is this book. *Sustainability* is admittedly something of a poorly defined buzzword. According to the Rio Declaration

(United Nations, 1992), promulgated by the international gathering in 1992 to examine global change issues, sustainable development will "equitably meet developmental and environmental needs of present and future generations." The term is commonly applied at a range of spatial scales, including the global scale (Sachs, 2015). However, the prospects for achieving sustainability at the global scale are as yet unclear.

BIG HISTORY

Given these key structural terms, let's consider the dynamics of the Earth as a whole, as an entity. The history of the geosphere, the biosphere, and the technosphere can be cast as a narrative. This chronicle includes the geological history of Earth, the biological evolution of life and the biosphere, and the cultural evolution of humanity. Thinking at the global scale thus forces us to juxtapose the astronomer's time frame of cosmic evolution (14 billion years), the geoscientist's time scale of Earth history (four billion years), and the historian's time frame of human development (approximately 10,000 years). The emerging field of "big history" has taken on this challenge (Chaisson, 2002; D. Christian, Brown, and Benjamin, 2013). The theme of increasing complexity is often used as the organizing framework in the big history literature, especially in academic courses designed to give students a cosmic perspective on the human condition. Increasing system complexity here is loosely defined as an increase in the number of interacting, self-organizing components within the system, often operating over a broadening range of spatial and temporal scales. In the context of global environmental change, we are practically forced to embrace the complexity paradigm because the problems range across multiple disciplines, notably including the biophysical sciences, social sciences, and, indeed, philosophy. This approach is increasingly referred to as "transdisciplinary." A question we must ask is whether humanity is capable of extending the evolution of complexity to a new level—that of a sustainable high-technology planetary civilization.

In the parlance of postmodernism, there is a "grand narrative" covering the origin and evolution of Western civilization. This model features the human conquest of Nature and the steady progress of technology and social organization toward our current glorified state. This grand narrative is, of course, now questioned for many reasons, not least among them the nascent threat that humanity poses to its own life support system.

We might thus pose a new master narrative for humanity, one more attuned to the planetary scale. Indeed, this new narrative—the Anthropocene narrative—provides a framework for this book (table 1.1). It begins with evolution of a self-organizing Gaian (if you will) biosphere. The next stage is the evolution of humans and their separation from Nature (i.e., the Great Separation). Through the acquisition of language, the taming of fire, the invention of simple tools, and the development of agriculture, we came to view Nature in mostly instrumental terms, and set ourselves

TABLE 1.1 Phases of the Anthropocene Narrative

NAME	DISTINGUISHING FEATURES
Evolution of a Gaian Biosphere (prehuman geological period)	Establishment of the global biogeochemical cycles that contribute to regulation of Earth's climate and the chemistry of its atmosphere and ocean
Great Separation (early human history)	Biological evolution of language-using *Homo sapiens*, control of fire, invention of simple tools, and domestication of plants and animals
Building the Technosphere (world history period)	Urbanization, construction of global transportation and communications networks, the Industrial Revolution
Great Acceleration (post–World War II)	Exponential increases in human population and use of natural resources, rapid advances in science and technology, and evidence of human impacts on the global environment
Great Transition (current)	Bending of the curves for global population size and natural resource use, emergence of a global awareness of limits on conversion of natural capital to technosphere capital
Equilibration (future)	Humanity learns to self-regulate and to manage the Earth system

apart from it. A subsequent Building of the Technosphere phase relied on gradual advances in technology and social organization, and produced global communications and transportation networks. The post–World War II Great Acceleration follows—an exponential phase of technosphere growth and influence on the Earth system. Up next may be the Great Transition; economist Kenneth Boulding (1964) and physicist Paul Raskin (2016) have employed this phrase to evoke a qualitative change in civilization in which rates of growth for global population and fossil fuel consumption would be stabilized and then reversed. Key to this concept is reconnection of the human enterprise with the biosphere. Lastly, we might imagine an Equilibration phase in which humanity learns to manage the biosphere and the global biogeochemical cycles—in effect, building a sustainable high-technology global civilization. As we shall see, there are no guarantees about fulfilling that aspiration (Ehrlich and Ehrlich 2012).

EARTH SYSTEM SCIENCE

The first scientific hint that the human technological enterprise had begun to have global-scale impacts on the environment came from observations of atmospheric CO_2 concentration in the late 1950s. During a postdoctoral fellowship at the Scripps Institute in San Diego, Charles David Keeling developed an instrument capable of reliably measuring the CO_2 molecule in a gas at very low concentrations. He then set out to study the spatial and temporal patterns in the atmospheric CO_2 concentration (technically, the CO_2 mixing ratio). At that time, there no way of estimating global fossil fuel emission of CO_2 or total CO_2 release from deforestation, so no one really knew to what degree they were influencing the CO_2 concentration in the atmosphere.

Keeling's first step was to set up a monitoring station high on the flanks of the Mauna Loa volcano on the Big Island of Hawaii. He reasoned that by situating his instrument in Hawaii, in the middle of the Pacific Ocean, local vegetation and fossil fuel sources would have a limited effect on the measurements. He was hoping to monitor the whole atmosphere from one sampling station. After taking daily measurements for a few years, Keeling made two major discoveries (figure 1.1). The first involved the recognition of a seasonal oscillation in the CO_2 concentration in the northern hemisphere, resulting in a decrease of about 10 ppm (parts per million; i.e., molecules of CO_2 per million molecules of dry air) in summer. That decrease turns out to be driven by a natural excess of carbon uptake through terrestrial photosynthesis (a carbon "sink") over plant residue decomposition (a carbon "source"). The biosphere could thus be said to breathe.

Keeling's second discovery was that the concentration of CO_2 in the atmosphere was rising. He didn't, by any means, know much about the various terms in the atmospheric carbon "budget," but it seemed likely to Keeling that the rise was caused by fossil fuel combustion. In 1957, geophysicist Roger Revelle famously characterized the rise in CO_2 as a global-scale experiment being conducted by humanity (Revelle and Suess, 1957). Scientists knew there might be impacts on the biosphere and climate system, but there was little certainty about how high CO_2 might go,

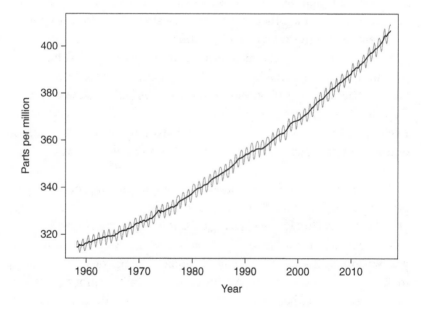

FIGURE 1.1

Historical record of the observations of CO_2 concentration at Mauna Loa. Adapted from NOAA (https://www.esrl.noaa.gov/gmd/ccgg/trends/full.html).

or the magnitude of the associated climate change. As we shall see in subsequent chapters, the sign of a human footprint on the atmosphere in 1955 was indicative of pervasive human impacts on the geosphere, hydrosphere, and biosphere.

We should step aside from the science of global change here to register what a truly profound discovery it was that humanity was altering the global atmosphere. Before the scientific era, the Western monotheistic religions had codified a creation myth that elevated humanity above the rest of creation. As retold in this myth, we had been given dominion over the Earth by an omnipotent deity and it was now our garden (i.e., the Great Separation). The Copernican revolution in the sixteenth century dealt a major blow to this belief system by establishing that Earth was not the center of the universe, but instead orbited the sun. This concept was contrary to Church dogma and significantly undermined Church authority. More than 200 years later, Darwin deduced that human origins were the same as any other animal's—i.e., we were the product of biological evolution. Through the lens of this theory, our facility with language was now viewed as just another adaptation, like sharp teeth for eating meat. This new perspective was rightfully humbling.

Keeling's discovery changed all that. It effectively placed humanity back into the mode of having dominion over the Earth. Whereas in earlier centuries that dominion was believed to have been bestowed on us by a creator God (in the Western tradition), geophysical observations by Keeling and others have made evident that we recently bestowed it on ourselves by the magnitude of our impacts on the planet (i.e., the Great Acceleration).

Today most people in both developed and developing countries have probably heard the idea that changes to the atmosphere are promoting global climate change. The big bang on this issue, in the United States, at least, seems to have occurred in the summer of 1988. The weather that year was unusually warm, and congressional hearings on climate change took place in midsummer in Washington, D.C., with temperatures near 100 degrees Fahrenheit outside the hearing room. Amid considerable debate and drama, there was a fever pitch of media arousal around the statement by climatologist James Hansen that the recent warming of the global climate could be attributed with high certainty to effects of rising greenhouse gas concentrations. On the cover of *Time* magazine, Planet Earth was depicted with a thermometer sticking out of it, and the temperature rising into the red zone. Congress was not ready to start limiting carbon emissions, but legislators took heed of scientists' warnings and made big commitments to more research, notably by funding an array of NASA's Earth Observing System satellites.

Earth system science subsequently emerged as a new scientific discipline that aimed to "capture all processes in nature and human societies as one interlinked system" (Lovbrand, Stripple, and Wiman, 2009, p. 12) at the global scale (Schellnhuber, 1999). Over the past 25 years, a burst of Earth system science research performed all over the planet has led in several directions, some of which point toward an ominous end (Clark et al., 2004). One of the core specialized fields in Earth system science is paleoclimate research. Through examination of the paleorecord (e.g., as found in ice cores and marine sediment cores), we have gained a clearer understanding of the role of greenhouse gases in Earth's climate over geological time scales (see chapter 2). There is no question that these gases have been dominant players in the dramatic swings in climate that have occurred in Earth's history. The associated environmental changes have had major consequences for life on Earth and are often drivers of extinction events. Examination of the paleorecord has also made it clear that the biosphere is an active influence on the atmospheric composition (see chapter 3). A key question for the purposes of our collective future is the degree to which the biosphere dampens or amplifies directional changes in the climate system. Specifically, could the biosphere compensate for human influences on the climate system? Or will it amplify the human influences and perhaps even precipitate a major extinction event?

A second focus in Earth system science has been on the current state of the climate and the global biogeochemical cycles (see chapters 4 and 5). Humanity was characterized by Vernadsky as a geological force primarily because it had begun to widely alter the cycles of carbon, nitrogen, and water. Scientists now use an array of global monitoring systems, ranging from networks of stream gauges to satellite-borne sensors of vegetation phenology, to track the human footprint on the global environment. The "green marble" is an image of Earth with plant productivity mapped by satellite remote sensing.

A third focal research area in Earth system science has been development of Earth system models that simulate the global biogeochemical
cycles and the global climate (see chapter 6). Business-as-usual scenarios
of CO_2 emissions and accumulation in the atmosphere point toward levels on the order of 500–1,000 ppm by 2100, up from approximately 280
ppm at the beginning of the Industrial Revolution. Earth system models
are tools for evaluating the climate consequences associated with those
concentrations. Despite variation among the models in their sensitivity
to greenhouse gas increase, there is now wide agreement among climate
modelers that anticipated increases in CO_2 and other greenhouse gases
(driven primarily by fossil fuel emissions) will profoundly alter global climate, sea level, and vegetation distribution. As more information and
theory are introduced into development of these models, the evidence is
increasing for a variety of positive (amplifying) feedback responses to the
initial human-induced greenhouse gas warming. The outcomes extend
from mostly manageable environmental changes, if emissions are rapidly
reduced, to doomsday scenarios, if business-as-usual emissions continue
and strong carbon cycle feedbacks are engaged. Remarkably, it appears
that humans have managed to divert upward the slow global cooling
trend evident in the paleorecord that was leading toward the next ice age
(figure 1.2).

The implications of improved understanding of Earth's climate system
and the human impact on it have inspired researchers to apply Earth system models in evaluating the prospects for climate change mitigation.
This field addresses both what can be done to reduce net greenhouse gas
emissions, e.g., create carbon sinks by afforestation (the planting of trees
in an area where they had not grown before), and what might be done to
minimize climate change impacts once the greenhouse gas concentrations have already risen. Thus far, geoengineering solutions such as
shading Earth by mechanical or chemical means are perceived as too
impractical or polluting. The danger of unintended consequences lurks
behind many of the technological fixes potentially applied to mitigate
human impacts on the environment, which highlights the importance of
a whole Earth modeling framework to evaluate them.

A last critical element of Earth system science is the study of the human
dimension of global environmental change (explored in chapters 7–10).

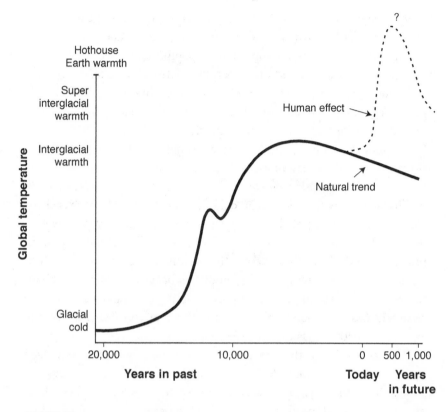

FIGURE 1.2

Humans have likely diverted the global climate trajectory from its path toward the next ice age. The background trend, driven by slowly changing solar forcings, was pushing Earth toward a return to glacial period temperatures. Anthropogenic greenhouse gases are now propelling the global mean temperature above the range of most previous interglacial periods, and toward the "Hothouse Earth" warmth of the mid-Cenozoic era over 30 million years ago. Adapted from Ruddiman (2005).

Here the concern is how people perceive and experience global environmental change, how they respond, and how we might organize ourselves for possible mitigation and adaptation. The social sciences come to the forefront as we seek to understand how people interact economically and politically at various levels of organization (Palsson et al., 2013). Ecological modernization theory (see chapter 7) suggests that global environmental change issues are addressable if ecological considerations rise to the level of economic and security considerations in all societal deliberations. The ecological sciences have developed a subdiscipline (anthroecology), which seeks to "to investigate and understand the ultimate causes, not just the consequences, of human transformation of the biosphere" (Ellis, 2015, p. 321). That project requires close attention to the history and prospects for linked socioecological systems (Angelstam et al., 2013).

German sociologist Ulrich Beck hypothesized that as global-scale threats such as nuclear holocaust and stratospheric ozone depletion become more prominent, universal exposure to the risks they represent will induce a new force for social cohesion (Beck, 1992). The risks at the global scale are so great that both rich and poor individuals, as well as nations, share vulnerability. Global environmental change may thus push humanity toward the kind of collective identity and global-scale institutions needed to manage ourselves and the global environment (see chapters 8 and 9). Certainly, it will require coupling of humans and natural resources (socioecological systems) at multiple levels of organization to arrive at something sustainable (see chapter 10).

Practitioners of Earth system science are, for the most part, motivated by a deep curiosity about the structure and function of the Earth system. Research scientists are generally supported by governmental funding, and the traditional social contract was that their research should produce an increasingly deeper understanding of how the world works. Earth system science is now in a position in which society is asking for more. Besides characterizing the impacts of humans on the Earth systems, practitioners of Earth system science must "work to provide the underpinnings for workable solutions at multiple scales of governance" (DeFries et al., 2012, p. 603). The policy community is in need of "actionable, sustainability-relevant knowledge" (Lubchenco, Barner, Cerny-Chipman, and Reimer, 2015). That thinking is increasingly reflected in the budget priorities of the major science funding agencies.

CONCLUSIONS

Our project of thinking globally is both an interdisciplinary and a transdisciplinary endeavor. The interdisciplinary part draws on fields in the geosciences such as paleoecology and climatology. These disciplines share a common foundation in physics, chemistry, and biology. The transdisciplinary part combines information and ways of knowing from the biophysical realm with knowledge from social science disciplines (especially sociology, economics, and political science). Here, the attribution of causality is more complicated and the capacity to model fundamental processes is more limited. Yet only by understanding both the biophysical and the human dimensions of global environmental change can we hope to design and operate a sustainable, high-technology, global civilization.

FURTHER READING

Schellnhuber, H. J., Crutzen, P. J., Clark, W. C., Claussen, M., and Held, H. (Eds.). (2004). *Earth system analysis for sustainability.* Cambridge, MA: MIT Press.

2

EARTH'S GEOSPHERE, BIOSPHERE, AND CLIMATE

We may now view earth history as a matter of evolution in which some changes are unidirectional (at least, in net effect), others are oscillatory or cyclic, and still others are random fluctuations, while the whole is punctuated by smaller or greater catastrophes.

—A. G. Fischer (1984)

The ability of Earth system scientists to retrace the history of Earth's geosphere, biosphere, and climate has grown remarkably in recent decades (Summerhayes, 2015). The rock record on land and in the ocean, as well as cores of glaciers and lakebeds, are probed with an increasingly sophisticated range of tools. The indicators of changing climate and biology include fossil species composition and abundance, isotope ratios in fossils and rock formations, and various other chemical signatures for processes and life forms. Earth system models (discussed in chapter 6) have been developed to simulate hypothesized environmental histories. The models make use of observational data for input as well as for validation. Results of geosciences research help characterize how much Earth's climate has varied in the past, how fast it changes, and by what mechanisms it is regulated. Understanding the past provides a route to understanding the probable future of our planet.

THE GEOSPHERE AND CLIMATE

Cosmologists largely agree about the origin of the universe—a "Big Bang" event that occurred about 14 Bya (billion years ago). Physicists have determined that even minor differences in the values of the physical constants (e.g., the strength of gravity and the forces that hold atoms and molecules together) would generally have made the interesting kind of universe we see around us impossible (Barrow, 2002). This appearance of design tends to make reductionist scientists feel a bit queasy, but it suggests in any case that we live in an order-friendly universe (Kauffman, 1995). That point is worth making in the context of the human aspiration to build a sustainable high-technology planetary civilization, mentioned earlier in chapter 1.

The history of the known universe can be traced back to a sea of hydrogen and helium, the simplest atoms. That sea became less homogeneous as it expanded, and gradually an assortment of astronomical entities— notably galaxies, stars, and planets—coalesced under the force of gravity. Stars are of special interest with respect to the origin of life because they are necessary to generate the higher atomic weight atoms that are necessary for life. The pressure of gravity in stars initiates nuclear fusion reactions in which atoms of the lighter weight elements are progressively fused into heavier atomic weight elements such as carbon, nitrogen, and phosphorus. The lifetime of each star is determined by the amount of fuel it accumulated initially. At the end of its lifetime, residual material may be widely dispersed and ultimately reconstituted into new stars and planets.

By the time our sun and solar system formed, about 4.5 Bya, their material content had been recycled through several generations of stars. The fact that the higher atomic weight elements are gradually accumulating in the universe sets the stage for an increase in complexity (Chaisson, 2002). The chemistry of life as we know it requires these higher atomic weight elements; a fact that lends support to the idea that the universe, besides being order friendly, could be said (albeit unscientifically) to be life friendly as well (de Duve, 2002).

Fusion reactions in the sun are associated with a tremendous release of energy. This process was important in the early development of the biosphere because one of the important conditions for the maintenance of a planetary-scale living system was a steady flow of electromagnetic energy from the sun

The geosphere coalesced under the force of gravity from a ring of debris orbiting the sun. During its formation, heavier elements—particularly uranium—were differentially concentrated in its core. The ongoing radio-active decay of these heavy elements generates the heat that drives con-vective cells of molten rock, which cycle between the Earth's core and its surface. This movement powers the shifting of tectonic plates on the sur-face (i.e., continental drift). One effect of tectonic plate collisions is volca-nism, which is important in Earth's history because it provides a source of gases to the surface. That flux of gases contributed to formation and maintenance of the atmosphere.

Perhaps the most spectacular event in the early history of the geo-sphere was its collision with a planet-sized body about 4.4 Bya. The result was formation of Earth's moon. The current tilt of the Earth's axis of rota-tion is a legacy of that collision. The gravitational influence of the orbit-ing moon stabilizes the angle of Earth's tilt, which has contributed to stabilizing the variability of Earth's climate (Laskar, Joutel, and Robutel, 1993).

For the first billion years of Earth's existence, the "degassing" of the lithosphere (its outer solid layer) proceeded at a rapid rate, resulting in the formation of an atmosphere and a hydrosphere. The atmosphere is a strong influence on the Earth's energy budget (discussed next) and, hence, its climate. The early atmosphere probably had high concentrations of hydrogen and various greenhouse gases—most likely water vapor, carbon dioxide (CO_2; J. C. G. Walker, 1985), and possibly methane (Pavlov, Kast-ing, Brown, Rages, and Freedman, 2000). The greenhouse gases would have kept Earth's climate favorable for life despite a less radiant sun. The atmosphere is relatively well mixed, thus tending to spatially homoge-nize the global climate despite the differences in solar energy arriving at the different latitudes.

Controversy remains regarding the origin of the large quantity of water on Earth's surface. Some of this water resulted from degassing asso-ciated with volcanism, and some from collisions of Earth with ice ball com-ets. Because of its great thermal inertia, the abundant water on Earth's surface helps stabilize the surface energy balance over diurnal and seasonal cycles. The ocean component of the hydrosphere is not as well mixed as the atmosphere, but poleward transport of heat in ocean currents also helps to homogenize climate.

It is important to think of the geosphere as an entity in its own right. The volcanism and mountain building associated with movement of the tectonic plates is constantly reshaping the surface of the planet and altering the chemistry of the atmosphere. In what is known as the "rock cycle" (figure 2.1), molten material is brought to the surface where it cools and crystallizes into rocks (Gregor, Garrels, Mackenzie, and Maynard, 1988). Physical and chemical processes (mineral weathering) then act on the rocks to break them down into low molecular weight compounds (salts) that are washed out to the ocean by the hydrologic cycle. In the ocean, evaporation of water leaves behind the salts, and their concentrations rise to a level at which they precipitate (i.e., crystallize) as new rocks on the

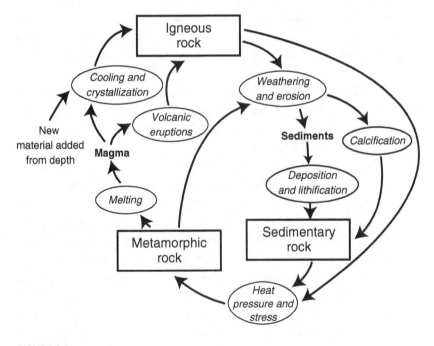

FIGURE 2.1

The rock cycle refers to cyclic transformations among the rock types.

ocean floor. These formations are folded into the crustal movements associated with plate tectonics and are thereby eventually returned to the land surface, where they are exposed to mineral weathering. This cycle takes hundreds of millions of years (Rampino and Caldeira, 1994).

We should not let the "uniformitarianism" associated with the image of the rock cycle blind us to the very slow trends in the overall development of the geosphere or to its episodic spasms of activity. The most important long-term trend is gradual decay of the radioactive material at the Earth's core. This fuel drives the movement of tectonic plates, but it will eventually be consumed. The slowing or cessation of tectonic activity, as has apparently already happened on Mars, will likely mean significant changes to Earth's atmosphere and climate. The erratic aspect of the geosphere shows up in repeated cases of massive igneous rock extrusion (i.e., flood basalt events) that have had profound impacts on the climate and biosphere.

EARTH'S ENERGY BALANCE

Earth's climate is ultimately a function of its energy balance (figure 2.2). The big players are the incoming radiation from the sun, the albedo (i.e., the degree to which the clouds and the surface reflect or absorb solar radiation), and the greenhouse gas concentrations in the atmosphere. The heat generated by decay of radioactive materials in Earth's core is not a significant direct influence on the surface energy budget.

With respect to incoming solar radiation, Earth is considered a "Goldilocks" planet (Rampino and Caldeira, 1994) because its distance from the sun is such that solar radiation is strong enough to prevent a frozen surface, but not so strong as to raise the surface temperature to the point of boiling water (thus restricting if not preventing life). Although scientists refer to the "solar constant," there has been a gradual increase (approximately 25 percent) in the amount of radiation reaching the Earth's surface from the sun over geological time (D. O. Gough, 1981). The change is associated with fusion reactions that cause contraction of the stellar core.

Once solar energy reaches the planet, it is either reflected or absorbed. If the Earth were simply a ball of black rock, it would absorb most solar

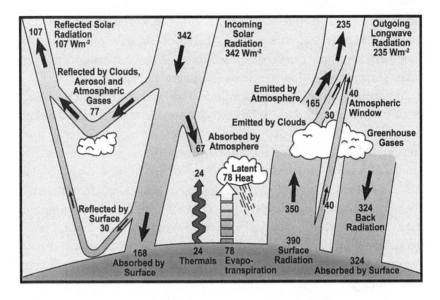

FIGURE 2.2

Estimate of the Earth's annual and global mean energy balance. Over the long term, the amount of incoming solar radiation absorbed by the Earth and atmosphere is balanced by the Earth and atmosphere releasing the same amount of outgoing long-wavelength radiation. About half of the incoming solar radiation is absorbed by the Earth's surface. This energy is transferred to the atmosphere by air warming in contact with the surface (thermals), by evapotranspiration, and by long-wavelength radiation that is absorbed by clouds and greenhouse gases. The atmosphere in turn radiates long-wavelength energy back to the Earth as well as out to space. Originally from Kiehl and Trenberth (1997). Figure and caption from FAQ 1.1, figure 1, IPCC chapter 1, "Historical Overview of Climate Change Science" (Le Treut and Somerville, 2007). Courtesy of the Intergovernmental Panel on Climate Change.

radiation. Eventually, energy equilibrium would be established between how much energy was absorbed and how much energy was reradiated. Generally, as a so-called black body absorbs more radiation, its temperature (reflecting the kinetic energy of its molecules) rises and it emits more radiation. All else being equal, the stronger the solar radiation reaching Earth, the higher its surface temperature would be. If our rock in space was white rather than black, meaning that it was made of molecules that reflected rather than absorbed solar radiation, the rock's temperature would be much cooler. Water on the surface of the planet adds one other wrinkle since absorbed energy can be dissipated in evaporation. Later condensation of the water vapor produced by evaporation warms the atmosphere.

The concentration of greenhouse gases is the third major factor in the Earth's climate system. These gases—notably CO_2 and water vapor—absorb some of the outgoing black body (longwave) radiation coming from the planetary surface and reradiate it at somewhat longer wavelengths, hence warming the planet. That additional longwave radiation is what constitutes the "greenhouse effect." Molecules vary as to the wavelengths they absorb and reradiate. Thus, they have differential effectiveness as greenhouse gases. The most important greenhouse gases in Earth's recent nonpolluted atmosphere, based on their concentrations and the wavelengths they absorb and reradiate, are water vapor, CO_2, methane, ozone, and nitrous oxide. The concentrations of these gases have varied widely over the course of Earth's history and in this chapter and the next we will attend to the controls on those concentrations.

Earlier we noted the rock cycle associated with the formation and dissolution of minerals. A key feature of the rock cycle in relation to the global climate is that it results in a crude negative feedback mechanism (box 2.1) for regulating global temperature (box figure 2.1). Over the course of Earth's history, CO_2 has been one of the most important greenhouse gases. If the CO_2 concentration in the atmosphere increases (e.g., because of increased volcanism), temperature generally warms. However, a high CO_2 concentration and warmer temperatures promote faster mineral weathering on the terrestrial surface. In the silicate weathering reaction, CO_2 from the atmosphere combines with water to form a weak acid that dissolves rock and results in carbon being transported to the ocean in stream flow as a bicarbonate ion. Consequently, there is a steady flux (a CO_2 sink) of CO_2 out of the atmosphere. Global warming can increase the reaction rates for the weathering sink and the flow of water, leading to a decrease in the atmospheric CO_2 concentration and climate cooling. Alternatively, when the CO_2 concentration decreases because of an imbalance in atmospheric inputs and outputs, the climate

BOX 2.1
Feedback and Homeostasis

The interactions among components within a system may be one-way cause-and-effect relationships or feedback loops. In the latter case, the component being stimulated to change does so in a way that weakens (negative feedback) or strengthens (positive feedback) the source of the stimulation. An example of negative feedback in an organism is the control of blood sugar in mammals. As the blood sugar level drops, the liver releases more glucose into the blood stream to bring blood sugar back up to a metabolic set point. If blood sugar gets too high, the opposite signal is sent. Hence negative feedback loops tend to be stabilizing. Positive feedback magnifies an initial

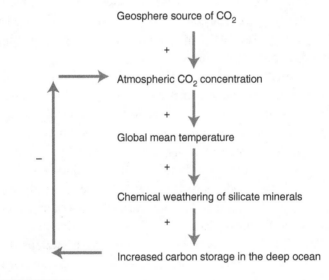

BOX FIGURE 2.1

The silicate rock weathering thermostat (Walker, Hays, and Kasting, 1981) operating as a negative feedback to CO_2 increase. The plus signs adjacent to arrows indicate that one process stimulates the next. A minus sign implies a suppressing effect. Increased carbon storage in the ocean reduces the atmospheric CO_2 concentration and moderates a global temperature increase.

change and is potentially destabilizing because increases are self-reinforcing and hence open ended. Infectious laughter between two people can build and build.

In Earth's climate system, there are many feedback loops (Lashof, DeAngelo, Saleska, and Harte, 1997). A simple case of negative feedback is the stimulation of photosynthesis by the rising atmospheric carbon dioxide (CO_2) concetration. More photosynthesis tends to draw CO_2 out of the atmosphere. A commonly cited positive feedback loop is the current greenhouse gas–induced climate warming, which reduces the amount of sea ice, which increases the surface absorption of sunlight, which further enhances the warming.

Homeostasis is the maintenance of a system at a reference condition. A home heating system is homeostatic in the sense that a decrease in temperature produces a signal to the furnace to turn on, which warms the air. When the air gets too warm, a signal is sent to turn off the furnace (a negative feedback loop). A current challenge to Earth system science is to understand the degree to which Earth's climate system is homeostatic and by what mechanisms homeostasis is achieved. Alternatively, what are the positive feedback loops that may amplify the current human-induced climate warming?

cools, the rate of weathering slows, and (given a continuous input of CO_2 from volcanism) the CO_2 concentration in the atmosphere increases and climate warms. This mechanism—the silicate rock weathering thermostat—is believed to underlie the unexpected stability in the Earth's climate over the past four billion years despite a significant increase in solar irradiance (J. C. G. Walker, Hays, and Kasting, 1981).

THE BIOSPHERE

Four billion years ago, Earth was lifeless. By 3 Bya, a vigorous layer of life—the biosphere—had established itself on the planet's surface. It was powerful enough to shape the chemistry of the atmosphere, the chemistry of the ocean, and the global energy balance. Hence, it had a persistent influence on the global climate. Biological evolution continuously shapes the composition of the biosphere and its biogeochemical cycling capacities. The geosphere, atmosphere, and biosphere change and influence each other over geological time.

Over the course of its development, the biosphere as a living entity has been vulnerable to three principal forces.

First is the sun. The biosphere depends primarily on solar radiation for its energy base (photosynthesis), but too much solar energy could excessively warm the global climate. As an entity in its own right, the sun has a history and a developmental trajectory over which the biosphere has no control. As noted, the amount of solar radiation reaching Earth each year has increased about 25 percent over last four billion years.

Second is the occasional collision with an asteroid or a comet. Such collisions have repeatedly damaged, and could potentially extinguish, the biosphere.

Third is the delivery of greenhouse gases to the surface from the geosphere. With too little gas, the Earth system could go into deep freeze; with too much, the biosphere could get overheated. What makes this issue immensely complicated is that the biosphere is a significant player in the biogeochemical cycles that control greenhouse gas concentrations.

A question we are particularly interested in here concerns the degree to which the biosphere can influence Earth's climate. Do its emergent properties stretch to include homeostatic regulation of climate in the face of increasing solar luminosity? Is it involved in negative feedbacks in the climate system that stabilize the climate, or mostly positive feedbacks that amplify climate change? This question is, of course, not purely of academic interest considering the expanding human pressure on the climate and biosphere. If the biosphere/geosphere/atmosphere system is homeostatic with respect to climate variation (discussed in chapter 3), might humanity somehow be saved from inordinately rapid global warming induced by anthropogenic greenhouse gas emissions? Likewise, if the biosphere is important to the functioning of the Earth's climate system, and is sensitive to human depredations, will further degradation make the human predicament even worse? Or more optimistically, could the biosphere be manipulated to ameliorate human influences on the climate system?

THE ORIGIN OF LIFE

Given the presence of a water medium, a sufficient concentration of various lower atomic weight elements, and the kind of energy sources that were available early in the history of the Earth, a variety of molecules that constitute the building blocks of early life forms were spontaneously synthesized by purely inorganic processes. Complex molecules of this type are even found in meteorites, suggesting they can be synthesized under the quite harsh conditions of deep space. Scientists can recreate the inorganic synthesis of these complicated molecules in the laboratory but cannot yet replicate the subsequent self-organization of life from nonliving chemicals.

Whole scientific journals are devoted to theories about how the most primitive life forms evolved. Here I will refer you to Smil's (2002) discussion and simply suggest there are good reasons to believe that under a regime of a steady flow of energy (solar or possibly chemical energy in undersea thermal vents) and the availability of suitable chemical building blocks, a series of steps was initiated early in the Earth's history that led to the evolution of simple living cells. Microbiologist Christian de Duve (2002) describes life as a "cosmic imperative" in that the universe is characterized by an incredible array of physical and chemical features that seem to underlie life's origin and maintenance. Biochemist (and sociologist!) Lawrence Henderson proposed a life-friendly universe in the early twentieth century and "Henderson's central idea—that the physical universe is intriguingly biofriendly—has endured through the revolutions in cosmology, physics, molecular biology, and astrobiology that followed in the decades after his work was published" (Davies, 2008, p. 111).

Life shows up in the fossil record about 3.8 Bya, soon after the end of a period of heavy asteroid bombardment that characterized the first billion years of Earth's existence. It is quite possible that life originated and was extinguished on several occasions before it became firmly established. Many scientists have speculated that life evolved independently on Mars and we may yet encounter it there as we make increasingly sophisticated attempts to find it. The relatively new discipline of astrobiology is predicated on the likelihood of life elsewhere in the universe.

Of course, even the simplest living cells represented a vast increase in order over the primordial chemical soup of inorganic molecules. However, an increase in order is predictable under certain conditions (Prigogine and Stengers, 1984). There was first the propensity for self-organization, seemingly built into the inorganic atoms and molecules already present (Kauffman, 1993). Membrane-like structures and even nonliving cell-like structures form spontaneously in the kind of chemical soup that is believed to have been present about the time life originated. Second, there were very powerful, continuous flows of energy to the surface of the Earth from the Sun and from geothermal sources. These energy sources drove the synthesis of the relatively complex molecules that became the components of, or the fuel for, the earliest forms of life. Third, although Earth is an open system with respect to energy, it is largely a closed system with respect to mass. Occasional comets and asteroids enter Earth's atmosphere and add small amounts of mass, but the chemistry at the surface is constrained by the chemical composition of the Earth's crust and outputs from volcanism. This constraint has implications for the early evolution of life because primitive microorganisms would have sequestered specific, relatively rare elements in their biomass until these nutrient elements became limiting. Thus, competition for limited resources became an issue. With reproduction by simple fission and competition for limited resources, we begin to see the basis for natural selection and biological evolution.

Early on, a means for transmitting information across generations about self-construction and self-maintenance of cells evolved. The molecules involved were RNA (ribonucleic acid) and DNA (deoxyribonucleic acid). Errors, or mutations, in this genetic material could then provide the requisite variation that is the basis for biological evolution. We can think of early biological evolution as a constant search for, and refinement of, biochemical pathways that allowed more energy flow through living entities.

THE BIOGEOCHEMICAL CYCLES

The earliest organisms were chemoautotrophic, meaning they derived their energy by uptake and breakdown of energy-rich inorganic chemical compounds. The accumulation of energy-rich organic molecules associated with these microbes, and the associated shortage of critical elements needed for metabolism, set the stage for evolution of primitive heterotrophic cells (i.e., cells that could digest other living or dead cells). Eventually, with consumer organisms that could ingest living cells, with autotrophs, and with decomposing heterotrophs, the basis for nutrient cycling (i.e., ecosystems) was achieved.

Given the limited availability of certain chemicals, the early proliferation of life could not have proceeded without nutrient cycling. Typically, a heterotrophic organism is most interested in the energy from the molecules that it consumes. It will release excess nutrients, such as nitrogen, as waste. Autotrophs then consume (recycle) that nitrogen. Commercially available "biospheres" are self-enclosed glass orbs that contain the right mixture of life forms to continuously recycle critical nutrients. If exposed to sunlight, these miniature ecosystems can maintain themselves indefinitely.

At the scale of the planet, we refer to the global biogeochemical cycles. Earth system scientists have identified pathways of the cycling of water, carbon, nitrogen, sulfur, and other elements critical to life. These pathways may include gas, liquid, and solid chemical states. Some steps involve living material and some do not. The energy driving the global biogeochemical cycles mostly comes from the sun, either directly—as in evaporation of water—or indirectly as in photosynthesis. We'll examine these cycles more closely in chapter 4 and evaluate how they have been altered by the technosphere.

THE EARLY EARTH SYSTEM (3.7–0.541 BYA)

Besides chemoautotrophy, which was initially an anaerobic (without oxygen) process, other significant metabolic capabilities evolved in the first billion years or so of life. Anaerobic photosynthetic bacteria (photoautotrophs) tapped into a vastly greater energy supply (the sun) than was available to the chemoautotrophs. In photosynthesis, chlorophyll molecules absorb sunlight; the electromagnetic energy is then used to construct chemical bonds in energy storage compounds. Carbohydrates (simple sugars derived from the products of photosynthesis) are a key biotic energy source when metabolized during respiration. The carbon atoms that form the backbone of photosynthetically produced carbohydrates come from atmospheric CO_2.

Initially, neither the photosynthetic pathway nor the respiratory pathway involved oxygen. Geologically produced compounds such as hydrogen sulfide (H_2S) provided the required hydrogen and electrons needed for photosynthesis (Smil, 2002). However, geological production of these reactants was limited, and eventually a chemical pathway evolved based on splitting a water molecule (H_2O) to gain hydrogen and electrons. The oxygen atom was released to form gaseous oxygen. Oxygen thus became a by-product of photosynthesis and could accumulate in the atmosphere.

Once photosynthesis and nutrient cycling were in place, the biosphere gained a significant degree of autonomy from the geosphere. It was no longer quite so dependent on the geosphere for a steady supply of energy and nutrients. The stage was set for a true biosphere (i.e., a functional entity as opposed to a spatially defined collection of organisms).

One of the first signs that a true biosphere had formed is evidence in the paleorecord for atmospheric oxygen (approximately 2.7 Bya). Oxygen (O_2) and its byproducts are very reactive molecules. Initially, reactions with minerals—principally iron in the oceans or reduced compounds released by submarine volcanoes (Kump and Barley, 2007)—consumed the oxygen produced by aerobic photosynthesis. Eventually, however, the supply of oxygen from photosynthesis overwhelmed the sink from reactions associated with those minerals, and the concentration of oxygen in the atmosphere began to rise.

This change in the environment (the so-called Great Oxygenation) was a global crisis from the perspective of the biosphere. Being so reactive, oxygen by-products tended to be toxic to the life forms that had evolved to that point under anaerobic conditions. The widespread diminishment of anaerobic life forms about this time could be considered the first great extinction event in Earth's history. Remarkably, it was induced by the biosphere itself.

However, the chemical nature of oxygen meant that it could potentially be used in chemical reactions that were favorable to life. As the oxygen concentration increased, life forms evolved that could exploit its chemical potential in a new form of respiration. In aerobic respiration, one glucose molecule (a simple sugar) produces 19 times more available chemical energy than occurs through the more primitive anaerobic respiration of that molecule. The increased energy availability set the stage for the evolution of more complex multicellular life forms. Meanwhile, primitive microbes that could not tolerate oxygen retreated to anoxic environments like deep ocean sediments (where they continue to thrive to this day).

Oxygen in the atmosphere also set the stage for the colonization of land. When short, high-energy, ultraviolet (UV) wavelengths of solar radiation strike an O_2 molecule in the upper atmosphere, an electron is knocked off, and subsequent reactions with other energetic molecules result in the formation of ozone (O_3). Ozone provides a significant benefit to land-dwelling organisms because it absorbs the UV wavelengths of solar radiation that can damage organic molecules. Prior to the buildup of ozone in the atmosphere, UV radiation from the sun reached the surface of the planet and was strong enough to disrupt metabolism—particularly the transcription of genetic information, which is critical for passing information (genes) from one generation to the next. Before the formation of ozone in the stratosphere, life was largely restricted to water bodies where UV radiation is attenuated with depth. After the ozone layer formed, life swarmed over the land surface.

Certainly by 2.2 Bya there was a functional biosphere—an entity with enough metabolic force to influence the atmosphere in significant ways. The biosphere not only drove up the oxygen concentration in the atmosphere but also drove down the concentration of CO_2 (as explained below). The concentration of methane (CH_4) may have also fallen at the same time. As noted, methane is a much stronger greenhouse gas than CO_2 and may (this point is still controversial) have been at high concentrations in the early atmosphere. The source of the methane would have been both volcanism and microbial metabolism.

Oxygen reacts with methane to form carbon monoxide (CO, not a greenhouse gas) which eventually oxidizes to CO_2. More biosphere-produced oxygen would mean less methane and hence, climate cooling. The mechanism of CO_2 drawdown mediated by the biosphere required enhancement of several steps within the global carbon cycle. Earlier we looked at the rock cycle (see figure 2.1), and now we can examine its relationship to the carbon cycle. As noted, the primary source of CO_2 to the

primitive atmosphere was volcanism, and the primary sink was formation of inorganic carbonate rocks in ocean basins. The shifting of tectonic
plates guaranteed a significant amount of volcanism and a steady recycling
of carbon buried in carbonate rocks back to the atmosphere.

The biosphere began to influence the ocean carbon sink by way of two
mechanisms, one in the ocean and one on land. In the ocean, phytoplankton absorbed CO_2 during photosynthesis to form biomass and took
up carbon during formation of calcium carbonate shells (e.g., diatoms).
Some fraction of the biomass and carbonate shells eventually settled to
the anaerobic ocean floor and accumulated in carbon-rich sediments.
This mechanism is referred to as the biological pump. On the land, the
mineral weathering reactions (i.e., the carbon sink of the rock weathering thermostat) involves weak acids attacking the crystalline structure
of minerals that make up silicate rock. Plant roots produce such acids
directly and, during root respiration, emit CO_2, which can react with
water to form an additional weak acid.

These acids effectively speed up the rate of silicate rock weathering,
and hence contribute to drawing down the atmospheric CO_2 concentration. As the biosphere extended to land, and when relatively large areas
of mineralizable rocks were exposed on the land surface, Earth's climate
began to cool.

Once global cooling began in the early history of the Earth, a strong
amplification (positive feedback) was driven by the water vapor content
of the atmosphere. Water vapor is an important greenhouse gas (in the
contemporary atmosphere it accounts for the largest proportion of the
greenhouse effect among all the greenhouse gases), and cooler air has less
water-holding capacity than warmer air. Simply put, cooling air temperatures mean less water vapor and hence less greenhouse warming.

Multiple episodes of "Snowball Earth" beginning about 2.3 Bya almost
certainly resulted from the combination of the relatively weak sun at that
time, a biologically driven decrease in greenhouse gas concentrations,
and the water vapor feedback (Melezhik, 2006). Once ice and snow began
to form on the planet surface, a new positive feedback mechanism to
cooling was engaged: the snow/ice albedo feedback. This purely physical
mechanism is based on the fact that open water and the land surface
largely absorb solar radiation, whereas snow and ice largely reflect it.

When temperatures cool to the point where sea ice and glaciers on land grow, the amount of solar energy absorbed at the planet surface decreases and global mean temperature drops further.

The term *Snowball Earth* refers to the idea that snow and ice entirely covered the surface of the Earth, probably even the low latitude oceans. These cold periods could last for millions of years. The escapes from Snowball Earth conditions were apparently extremely fast. Collisions with asteroids may have initiated the periodic escapes by melting the ice and snow cover and possibly causing catastrophic releases of accumulated greenhouse gases. Geological records from this period indicate carbonate deposits on top of glacial till; the carbonate would be from a burst of mineral weathering driven by a very high CO_2 concentration.

Earth was in the grip of another long cold spell from around 0.7 Bya, possibly associated with the ongoing redistribution of the continents (i.e., movement of the tectonic plates). The geographic arrangement of the continents influences the circulation of the atmosphere and the oceans, and hence the global energy balance. As noted, the distribution of land also determines the amount of easily weatherable rock that is exposed on the surface (Smil, 2002). Over geological time, that factor has varied depending on the amount of surface area above sea level and its mineral makeup.

The large swings in global climate in the early history of the biosphere seem to imply that the biosphere was not capable of strong negative feedback responses to global cooling. Its primary effect was to cool the climate by exerting downward pressure on the CO_2 concentration by way of its influence on the rate of mineral weathering.

Schwartzman (1999) has proposed an interesting link between the biosphere's cooling effect on the early global climate and the general trend in early evolution toward increasingly complex life forms. He hypothesized that high global mean temperatures in Earth's early history placed a limit on the complexity of biological life forms. In examining the temperature tolerances of various extant life forms, he found that a sequence from tolerance of high temperature to tolerance of moderate temperature parallels the sequence in which these organisms emerged in evolutionary history. The sequence was roughly cyanobacteria (prokaryotes), anaerobic eukaryotes, and then primitive multicellular organisms. The implication is that as soon as the climate cooled sufficiently to permit the evolution of more complex life forms, these forms evolved. By increasing rates of carbon uptake and burial by way of the carbon sink component of the silicate rock weathering thermostat, the new life forms had the effect of drawing down the concentration of CO_2—a sequence suggesting a positive feedback between the biosphere's influence on the climate and the climate's influence on the evolution of the biosphere (more about that in chapter 3).

THE MATURE EARTH SYSTEM (541 MYA–PRESENT, THE PHANEROZOIC EON)

The limits of the natural envelope of Earth's climate system for the bulk of the Phanerozoic time ranged from 180 ppm CO_2 and an average global temperature of around 11°C in peak glaciation conditions at the low end to somewhere between 4500 and 8500 ppm CO_2 and 30–32°C in peak hothouse conditions at the high end.

—C. P. Summerhayes (2015)

After 500 Mya (million years ago), the big story in terms of the biosphere was the conquest of the land. Primitive plant-like organisms such as lichens had colonized the land surface before 1 Bya, but by 350 Mya complex plants and animals fully occupied the land surface. The layer of vascular plants on the land surface began to influence the climate significantly. Notably, transpiration (i.e., evaporation from leaf pores) returns moisture to the atmosphere, which allows for deeper penetration of moisture to the continental interiors as air masses move inland. Transpiration also alters the surface energy balance, tending to cool the land surface temperature. Animals speed up the carbon cycle by digesting plant biomass. Generally, biological evolution can be viewed as a series of innovations that gradually add complexity to biosphere structure and function.

A new sphere was formed about this time: the pedosphere ("ped" is a Latin root word meaning "foot," which in this context alludes to soil, the surface we walk upon). The soil is a matrix of mineral particles, dead organic matter, and living organisms that provides a substrate for land-dwelling plants. Soil is beneficial to plants because of its high water-holding capacity and as a reservoir of nutrients that are slowly released through the decomposition process. Long after the use of solar energy in photosynthesis helped create the biosphere, the soil nutrient reservoir helped make the terrestrial biosphere more independent from the geosphere. Soil is important with regard to the biosphere's influence on the climate because it greatly increases the surface area of mineral particles subject to weathering, hence enhancing the sink for CO_2.

For most of the past 500 million years, Earth's climate was warmer than its current state (Royer, Berner, Montanez, Tabor, and Beerling, 2004). Atmospheric CO_2 was generally greater than 500 ppm (parts per million) (Franks et al., 2014; Retallack, 2002), and often greater than 1,000 ppm (compared with 180–280 ppm in the past three million years). Conditions were usually favorable for life in the oceans and on the land surface. Biodiversity proliferated. Based on proxies for temperature and CO_2 concentration, there was a consistent coupling of greenhouse gas concentrations and global climate (Fletcher, Brentnall, Anderson, Berner, and Beerling, 2008). However, the long intervals of benevolent climate were interrupted by occasional episodes of high CO_2, extreme climate, and altered chemical states (Retallack, 2009).

Oxygen concentration varied between 10 and 35 percent during this period (relative to the current 21 percent). Although arthropods (organisms with an exoskeleton) and vertebrates had successfully colonized the land by 360 Mya, an episode of low atmospheric oxygen (Ward, Labandeira, Laurin, and Berner, 2006) induced an extinction event about that time (the so-called Romer gap in the fossil record). The cause of the low-oxygen episode is uncertain, but it speaks to a limited capacity for biosphere-driven homeostasis with respect to oxygen. An evolutionary resurgence began around 345 Mya, and a sustained period of high CO_2 warmth began after 300 Mya.

Two additional major extinction events are evident in the fossil record over the next 200 million years. In the end-Permian mass extinction (EPME) event (about 252 Mya), a prolonged episode of flood basalt flow occurred. These plumes of molten rock reaching the Earth's surface drove up the CO_2 and methane concentrations. Various interpretations of what caused the mass extinction have been offered, but here is one scenario (Knoll, Bambach, Payne, Pruss, and Fischer, 2007).

1. The event starts with an initial pulse of CO_2 from the geosphere that significantly warms the atmosphere and ocean. An increase in atmospheric methane also seems to be involved, possibly induced by interaction of the flood basalt with carbon-rich rock formation (Retallack and Jahren, 2008). Alternatively, the warming of the ocean initiates melting of a large reservoir of frozen methane hydrates on the ocean floor. As noted earlier, methane is a product of microbial metabolism and under conditions of high pressure and low temperature in the deep ocean it can bond with water molecules and be sequestered in the form of frozen methane hydrates. Once released by warming, that methane diffuses into the upper ocean and eventually the atmosphere, with the effect of amplifying the warming.

2. The high global warmth leads to intense stratification of the ocean. It also causes the rate of mineral weathering to increase. The combination of high productivity in the ocean from the weathering-induced pulse of nutrients and associated consumption of oxygen by respiration of the organic matter results in a shift in the chemistry of decomposition. A key end product is hydrogen sulfide.

3. Hydrogen sulfide is toxic to most organisms, and its accumulation in the stratified ocean causes a wave of extinction among marine invertebrate organisms. Green sulfur bacteria, which use hydrogen sulfide in their photosynthesis metabolism, become prominent. The hydrogen sulfide also diffuses into the atmosphere and causes extinction among terrestrial species both directly, and because it reacts with ozone, thus dissipating the biosphere's shield from solar UV radiation.

4. The escape from these conditions comes from a decline in the geological source of CO_2 and a gradual sequestration of excess atmospheric carbon by way of the rock weathering thermostat (box figure 2.1).

Geologists continue to debate the relative magnitude and sequencing of these events (Cui and Kump, 2015; Schobben, Stebbins, Ghaderi, Strauss, Korn, and Korte, 2015). In any case, it appears that elevated greenhouse gas concentrations from the geosphere radically warmed the climate and that the biosphere itself contributed to the process of mass extinction (i.e., with hydrogen sulfide).

The other major extinction event of the period occurred at 65 Mya. In this case, there is good evidence that a relatively large asteroid collided with Earth. The collision would have filled the atmosphere with enough dust and gases to radically change the climate and the amount of sunlight reaching the surface. A surge in volcanism and associated climate warming before the asteroid impact apparently contributed to stress on the biosphere (Petersen, Dutton, and Lohmann, 2016). Approximately half the extant species of land plants and animals (notably many dinosaur species) went extinct.

After each of these extinction events, the biosphere gradually recovered and the geosphere was recloaked with a new array of plant and animal species. In some ways, extinction events reset evolution. Surviving species provide the source for new evolutionary lineages. Famously, the demise of the dinosaurs around 65 Mya paved the way for the ascendency of mammals.

Another conspicuous event in the geological record that indicates instability in the climate system and the linkage of CO_2 and climate (Zachos, Dickens, and Zeebe, 2008) is the Paleocene–Eocene thermal maximum (PETM). This 4°C–6°C spike in Earth's temperature around 55.5 Mya is termed a *hyperthermal* because it occurred during a period when Earth was already quite warm. Temperate forests were (already) growing around what is now the Arctic Ocean (Maxbauer, Royer, and LePage, 2014). The cause of the climate change may have been similar to the cause of the EPME event, i.e., a geosphere-driven spasm of CO_2 and methane emissions. CO_2 may have increased to more than 4,000 ppm. The effects were not as extreme as in the EPME, but the ocean acidified, deep ocean carbonates dissolved, and marine extinctions occurred (Kump, 2011). This warm episode lasted about 100,000 years. As the sources of methane and CO_2 eventually declined, the continuous production of oxygen by the biosphere reduced the methane concentration while continued terrestrial and marine sinks for CO_2 pulled down the CO_2 concentration.

From about 50 Mya onward to the present, Earth's climate has cooled (figure 2.3). Within this cooling trend, global mean temperature has had its ups and downs, including the recent ice age cycles over the past three

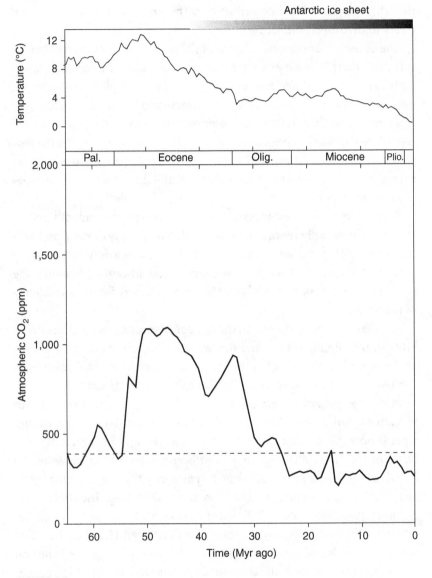

FIGURE 2.3

Indicators of Earth's climate and atmospheric chemistry over the last 60 million years. The CO_2 record is a central trend among multiple indicators. Adapted from Beerling and Royer (2011).

million years. But a cooling trend is clearly evident and the Earth system made a transition from a relatively warm to a relatively cold state. The geophysical basis for the cooling trend was largely a decrease in the greenhouse gas concentrations, primarily CO_2. Positive feedback to cooling by way of more ice and snow contributed. The cause of the CO_2 decline is still under debate, but the most widely supported hypothesis focuses on increased availability of readily weatherable rock associated with the uplift of the Himalayan Mountains (Raymo and Ruddiman, 1992).

The general trend of decreasing CO_2 since about 50 Mya was probably a significant stress to plants that had evolved at much higher CO_2 concentrations. One indication of a CO_2 shortage is an increase in the density of stomata on the surface of plant leaves in the fossil record (Beerling and Woodward, 1997). Stomata are microscopic portals on the leaf surface that allow diffusion of CO_2 (required for photosynthesis) into the leaf. When atmospheric CO_2 is high, stomatal density is low, so loss of water through the stomata is minimized. Low CO_2 levels are associated with high stomatal density.

CO_2 levels below approximately 200 ppm evidently drove the evolution of an alternative biochemical photosynthetic pathway, C4 photosynthesis (Sage and Monson, 1998). The C4 pathway is more efficient that the older C3 photosynthesis at low CO_2 concentration. It consists of anatomical and biochemical adaptations that essentially pump CO_2 into the leaf. It is now widely spread throughout the plant kingdom, which suggests that it evolved independently in multiple lines of descent.

Interestingly, the biosphere may effectively set a lower limit on the CO_2 concentration (and hence planetary cooling). Under low CO_2 conditions, the biosphere becomes CO_2 starved and its influence as a facilitator of mineral weathering (a CO_2 sink) is reduced. This mechanism would serve as a negative feedback mechanism to CO_2 decrease. However, the strength of this negative feedback is not well understood.

THE ICE AGE CYCLES (3 MYA–PRESENT)

Beginning about 3 Mya, Earth's climate began oscillating irregularly between warm and cold periods. Budyko (1984) proposed that the long-term trend of decreasing CO_2 concentration eventually cooled the global climate below a threshold such that polar ice caps became a continuous feature of the planet's energy balance. Recent modeling supports a threshold of about 600 ppm of CO_2 (Galeotti et al., 2016). Changes in paleogeography were also a factor in glaciations, though not dominant (DeConto and Pollard, 2003). Around 30 Mya, the Drake Passage between South America and the Antarctic opened and a circum-Antarctica current became established that isolated the Antarctic from the warmer air and sea water of lower latitudes. This new configuration supported the development of the Antarctic ice sheet. Around 3 Mya, the Isthmus of Panama closed, which changed ocean circulation in a way that favored glaciation on Greenland (Bartoli et al., 2005).

Over the past 800,000 years, the glacial/interglacial oscillations have become quite large and regular in duration, with a cycling time of about 100,000 years. Extended cool periods lasting most of the 100,000 years are flowed by relatively brief warm periods. In the early 1900s, Serbian civil engineer and geophysicist Milutin Milankovič (mi-LAN-kuh-vitch; 1879–1958) suggested that these cycles were associated with slight variations in the latitudinal distribution of solar energy reaching the planet's surface. These variations (termed *forcings*) are the result of temporal changes in the geometry of Earth's relationship to the sun. They include the variation in the elliptical orbit (a 100,000-year cycle), variation in the angle of the tilt (a 40,000-year cycle), and variation in the direction of the tilt (a 22,000-year cycle). The Milankovitch cycles run independently of one another, so their effects are sometimes additive and sometimes cancel out. The complexity of the interactions among the cycles and the slow feedbacks in the global climate system have made it difficult to precisely model their effects (Summerhayes, 2015).

To explain the sensitivity of the climate system to the Milankovitch forcings, we must look again at feedback mechanisms in the climate system. Recall that with positive feedback the response of a system to a perturbation tends to amplify the perturbation, whereas with negative feedback the system responds in a way that suppresses the source of the perturbation. Once sea ice and ice caps on land start growing under the influence of slightly reduced solar radiation to the northern hemisphere (or a difference in its seasonal distribution), the snow/ice albedo feedback kicks in,

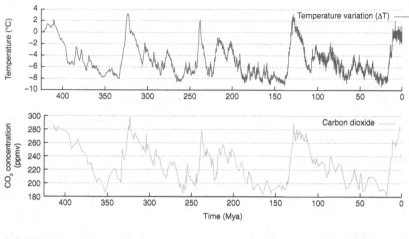

FIGURE 2.4

The paleorecord of atmospheric CO_2 concentration and Antarctic temperature over the last 400,000 years based on the ice core record. Mya = million years ago. Adapted from Petit et al. (1999).

the water vapor feedback kicks in, and, as we shall see, a CO_2/methane feedback kicks in. These are all positive feedback mechanisms. Milankovitch orbital forcings are thus the "pacemaker" of the recent glacial/interglacial cycles, but strong geophysical and biogeochemical feedbacks push the climate system to the extremes of cool glacial maxima and warm interglacial periods. The global mean temperature has varied by about 6°C over the course of the recent glacial/interglacial cycles.

The role of CO_2 concentration in the glacial/interglacial cycles is of particular interest with regard to understanding the relationship of the biosphere and global climate. The ice core record, i.e., the measurements of CO_2 concentration in air bubbles trapped in layers of snow that accumulate at high latitudes, shows that the CO_2 concentration has oscillated between about 180 and 260 ppm over at least the past 800,000 years (figure 2.4). Low values correlate with glacial maxima and high values correlate with the warmer interglacial periods. Like the CO_2 concentration, the temperature record over the glacial interglacial cycles is derived from ice cores, in this case the indicator is an isotope of oxygen whose concentration varies as function of temperature when the snow formed.

In terms of the physics of the global climate, this difference of 80 ppm accounts for about one third of the 6°C variation in global mean temperature over the glacial/interglacial cycle. The rest is driven by the Milankovitch forcings and the positive feedback loops in the climate system.

The 80 ppm increase in CO_2 concentration as the climate system comes out of a glacial maximum requires an increase of 160 billion tons (= 10^{15} grams [g] = 1 petagram [Pg]) of carbon in the atmosphere. The most likely sources of that carbon are the terrestrial biosphere, which has about 700 Pg of carbon in vegetation and 1,500 Pg in soil, and the ocean, which has about 38,000 Pg of carbon in an inorganic form. The change in terrestrial carbon storage during the transition from the cool glacial maxima to the warm glacial minima appears to be the opposite of what is needed to drive an atmospheric CO_2 increase: Carbon storage on the land is likely higher during warm periods because of the greater ice-free land surface (despite higher sea level) and wider distribution of forests. If the carbon pool increases in both the atmosphere and the terrestrial biosphere as climate warms, the ocean must be the source of that carbon. Modeling of the ocean carbon cycle as it responds to warming and retreat of polar ice supports that interpretation (Martinez-Boti et al., 2015). The mechanisms involve both a declining biological pump and changes in ocean circulation (Lourantou et al., 2010)

The ice core record reveals several abrupt climate changes that occurred as the planet was coming out of the last glacial period. These included the Younger Dryas event about 12,000 years ago that briefly returned the northern hemisphere to glacial maximum conditions. "Dryas" here refers to an alpine tundra wildflower, the pollen of which proliferates in the pollen record during cold periods. Rapid cooling events such as these are particularly informative when considering mechanisms driving climate change. A change in the intensity of the oceanic thermohaline circulation (THC) was apparently involved in this case. This global-scale vertical and horizontal circulation in the ocean (Broecker, 1987) generally results in warm water being driven north in the eastern Atlantic Ocean by the Gulf Stream, thereby warming the high latitudes in the northern hemisphere (Rahmstorf, 2003). Slowing of the THC, probably caused by a pulse of fresh water to the ocean from large glacial melt–fed lakes in North America, was most likely responsible for cooling at about the time of the Younger Dryas event. During deglaciation, rapid cooling events were not related to changes in greenhouse gas concentrations. The CO_2 concentration dipped with each cooling event, but not enough to be a strong positive feedback.

The peak of the current interglacial warm period occurred about six thousand years ago. At that time, the latitudinal limits of the tree line extended well north of their present location. Gradual cooling over the intervening millennia, driven by Milankovitch forcings, ended abruptly, just recently, when concentrations of all greenhouse gases began to rise, with an associated warming of the global mean temperature. The technosphere had begun to assert itself!

EARTH'S FUTURE

Budyko (1984) suggested that the higher CO_2 world of the near future associated with human greenhouse gas emissions will "rejuvenate" the biosphere. It will be a return to the warm, wet, high CO_2 conditions of the mid-Cenozoic era (30–50 Mya). As long as CO_2 remains high, the relatively weak Milankovitch forcings will not precipitate another ice age. Assuming that the pulse of anthropogenic CO_2 input tapers off over the next century, the CO_2 concentration may top out at somewhere between 500 and 1,000 ppm. It could reach 2,000 ppm if all readily available fossil carbon is burned or strong positive carbon cycle feedbacks are engaged (see chapter 6). The biosphere-assisted rock weathering thermostat and uptake by the ocean will then begin slowly drawing down the CO_2 concentration to what it had been before the anthropogenic perturbation, a process that will take thousands of years (Archer, 2010).

In the longer term, the steady increase in solar luminosity over geological time must be considered. Based on the mass of the sun and our understanding of the nuclear reactions that produce its radiation, astrophysicists predict that the trend of increasing solar radiation will continue for several billion years into the future.

Earth's temperature is, of course, strongly dependent on solar luminosity: All else being equal, the global temperature will increase in parallel with the sun's luminosity. As we have seen, there has been a reduction in the concentration of CO_2 over geological time and the reduction has, to some degree, compensated for increasing solar luminosity. As noted, the decrease in greenhouse gas concentrations appears to be in part a function of a sustained enhancement of the CO_2 sink from weathering by the terrestrial biosphere. However, the biosphere has done about all it can do as far as its cooling influence. In Earth's past, the CO_2 concentration decreased to levels that significantly reduced the rate of photosynthesis, thus lowering the vitality of the biosphere itself. But even if CO_2 is maintained in the range of 200–300 ppm (as was the case during the ice ages), the global mean temperature will eventually warm as solar luminosity increases.

As global mean temperature continues rising in response to aging of the sun, the biosphere will gradually simplify back to the thermophilic bacteria that will have persisted since its origin (Franck, Bounama, and von Bloh, 2006). Eventually, plate tectonics will cease and the planet will lose its water and atmosphere, as has already happened on Mars. As solar fuel approaches exhaustion, the sun's outer layer will expand and may ultimately engulf Earth.

ESSENTIAL POINTS

Cosmologists trace a path from the primordial sea of hydrogen, through formation of galaxies and stars, to the eventual formation of our home planet. Earth system scientists take over from there and describe the history of the geosphere, atmosphere, and biosphere. Each of the spheres is actively involved in regulation of Earth's climate. In the broadest view, greenhouse gas concentrations have fallen over the four-billion-year history of the planet, and Earth's climate has correspondingly cooled (despite increasing solar luminosity). This trend is driven by a strengthening CO_2 sink associated with enhanced weathering by the biosphere and increased land area available for weathering. The four-billion-year trend is punctuated by episodes of extreme cold and intensive warmth, in some cases associated with mass extinction events. The cold events were likely induced by biosphere-driven drawdown of greenhouse gas concentrations. The warm events were induced by geosphere sources of CO_2 and methane. The recent era of glacial/interglacial cycles emphasizes the importance of strong positive feedback loops in the climate system.

IMPLICATIONS

1. Earth's climate has varied widely over geological time, from eras of Snowball Earth to those of "Hothouse Earth," when temperate zone forests grew in what is now the Arctic zone. There is an emerging consensus among paleoclimatologists that global climate has been closely linked to greenhouse gas concentrations in the atmosphere throughout the history of the planet. This relationship strongly supports the conclusion that the ongoing human-induced increase in greenhouse gas concentrations will profoundly alter the global climate.

2. In Earth's history, volcanic eruptions and flood basalt events have repeatedly pushed greenhouse gas concentrations to high levels that have, in turn, drastically warmed the global climate, in some cases inducing extinction events. The human-driven combustion of fossil fuels is in many ways an analogue to these previous episodes.

3. The past three million years have been characterized by glacial/interglacial cycles, and the past 10,000 years are part of the warm interglacial portion of the most recent glacial/interglacial cycle. This period of relatively stable and warm climate has provided conditions favorable for the development of our advanced technological civilization. By injecting large quantities of greenhouse gases into the atmosphere, humanity has now initiated a super-interglacial period that will likely last thousands of years.

4. For much of the past 500 million years, the Earth system existed in a warmer, higher CO_2 condition than at present. Returning to that state will likely not be catastrophic for the biosphere. A problem for humanity may be the unprecedented rate of change (see chapter 6).

5. The history of Earth and its biosphere suggests that we live in an order-friendly and life-friendly universe. This is a sign that it may be possible to achieve the new level of order on Earth required for a sustainable, high-technology, global civilization.

FURTHER READING

Kump, L. R., Kasting, J. F., and Crane, R. G. (2010). *The Earth system*. San Francisco: Prentice Hall.

Smil, V. (2002). *The Earth's biosphere*. Cambridge, MA: MIT Press.

Summerhayes, C. P. (2015). *Earth's climate evolution*. West Sussex, UK: Wiley Blackwell.

3

THE EVOLUTION OF THE BIOSPHERE

What I see is that life produces byproducts and side effects that can shove the environment around into various chemical states. All organisms, linked in the chemical vessel of the biosphere, must adapt to these states or go extinct.

—T. Volk (1998)

As is evident from chapters 1 and 2, the field of Earth system science sprawls across many disciplinary boundaries. From its beginning, there has been a struggle to formulate unifying principles or narratives that are scientifically valid as well as broadly applicable and compelling. Perhaps the most well-known thrust in that direction is the Gaia hypothesis. Most broadly, it suggests that Earth as a whole is analogous to a living system, and correspondingly has homeostatic capability. The concept is attractive because it encompasses all the physical and biological properties of Earth, and proposes a basis for their integration as a system.

As we will see, the homeostatic Gaia concept has been rejected by geoscientists because of its teleological packaging. Nevertheless, the Gaia hypothesis matters to Earth system science because of its scientific implications: It has inspired researchers to look carefully at global-scale processes and at the mechanisms by which Earth's environment is regulated. It matters politically because whether or not there is Gaian homeostasis impacts the urgency of doing something about human-induced climate change. It matters to global environmental governance because the belief in a planetary-scale living entity has been adopted by many nonscientists and underlies their support for addressing key issues associated with global environmental governance.

THE GAIA HYPOTHESIS

The original Gaia hypothesis of James Lovelock and Lynn Margulis (expounded in the 1970s) held that the biosphere is part of a self-regulating biogeochemical cycling system on Earth that maintains conditions favorable to life (Lovelock 1979; Lovelock and Margulis, 1974). Two observations were key to their thinking. The first was that Earth's atmosphere is unusual among the atmospheres of planets in our solar system in having significant concentrations of reactive gases, including oxygen and methane. Oxygen is critical for aerobic respiration and is used in the production of metabolic energy by much of the biomass on Earth; methane is a greenhouse gas that significantly warms the global climate. Lovelock was a chemist and realized that since oxygen and methane readily react, their notably high concentrations in Earth's atmosphere could only be maintained by a high rate of production. The likely source of those gases was the biosphere.

The second observation was that climate conditions on Earth have remained within a habitable range (i.e., above the freezing point of water and below its boiling point) over the past four billion years. At first glance, climate stability may not seem remarkable. However, as we saw in the last chapter, that stability has been in the face of an approximately 25 percent increase in solar luminosity, a large variation in the rate of greenhouse gas inputs to the atmosphere from tectonic sources, and multiple catastrophic collisions with meteorites. Even if all else were constant, the increase in solar luminosity alone would have increased global mean temperature somewhere between 6°C and 25°C (Celsius), thus pushing the planet to more Venus-like conditions.

Lovelock considered the influence of the biosphere on the greenhouse gas concentrations as a key mechanism by which Gaia maintains homeostasis (Lovelock and Watson, 1981). And he challenged the geosciences community to work out other mechanisms. However, the reception of his ideas by the scientific world was generally not very positive.

One problem was that the original formulation of the Gaia hypothesis had a mythological tone to it: *Gaia* is the Greek name for the mother goddess. The name was suggested by Lovelock's neighbor, British author William Golding (*Lord of the Flies*). Lovelock sought to popularize the concept with a book in 1979 (*Gaia, a New Look at Life on Earth*). In doing so he gave Gaia a rather teleological tone, somewhat more philosophical that strictly scientific.

Another recurrent complaint was that the Gaia hypothesis was not testable. Classically, a scientific hypotheses or theory is "a statement about some phenomenon in nature that in principle can be confronted with reality and possibly falsified" (Bak, 1996, p. 162). Supporters of Gaian homeostasis thus have worked to identify feedback loops (see box 2.1 in the previous chapter) by which the biosphere interacts with the climate. A weak test would be whether biosphere feedbacks in the climate system are always negative (i.e., stabilizing). If they are positive, the biosphere would have to be considered disruptive rather than homeostatic.

The problem with testing specific predictions from Gaia theory is that we are just beginning to be able to monitor global-scale biosphere processes and trends (see chapter 9). We began directly monitoring global mean temperature only about 100 years ago and have monitored atmospheric carbon dioxide (CO_2) concentration only since 1958. The record of past states of the Earth system in the paleorecord (e.g., ice cores and the chemical properties of rock formations) can be used to evaluate feedback mechanisms to some degree, but these records are rather sparse, and interpretations are subject to controversy. To compliment observations, Earth system scientists have built simulation models of the Earth system to test their ideas about Gaian homeostasis. Here, too, there is much debate as to validity (Schellnhuber, 1999).

Despite the qualms within the scientific community, the American Geophysical Society (in response to advocacy by scientists including Carl Sagan and Steve Schneider) sponsored two Gaia-themed workshops (Schneider and Boston, 1991; Schneider, Miller, Crist, and Boston, 2004). The topic was popular around 1990 because the issue of human-induced global climate change was beginning to engage the geosciences community. The appeal to geoscientists lay in creating a multidisciplinary forum open to speculations about global-scale process that involved the biosphere. There was a general recognition that the biosphere indeed has a role in global climate regulation. But as we shall see, there was also substantive criticism of making general conclusions about Gaian homeostasis (Kirchner, 2003).

Besides its scientific legacy, the concept of Earth as Gaia has had a significant impact on the environmental movement. Lovelock originally was not attempting to inspire the environmental movement; in fact, his first book was rather hostile toward environmentalists (Lovelock, 1979). However, he did maintain that "As the transfer of power to our species proceeds, our responsibility for maintain planetary homeostasis grows with it" (p. 131). Many people in the environmental movement warmly embraced Gaia despite Lovelock because the concept not only got people thinking about global change issues but also became the platform for development of a global environmental ethic (Lautensach, 2008).

Because of the possible dire consequences for advanced technological civilization associated with human perturbation of Earth's climate system, there is now tremendous interest among the science, policy, and environmental communities in understanding global climate regulation. If Gaian negative feedbacks to climate change turn out to be strong enough, we might be saved from our own excesses. If the biosphere feedbacks to anthropogenic warming are predominantly positive, there is greater urgency to mitigating climate change. If humanity is the product of a benevolent global entity (Gaia), then perhaps we should learn to love and protect that entity (in a sense our Mother).

THE MEDEA HYPOTHESIS

In a pointedly contrary view to the Gaia hypothesis, Professor of Earth Science Peter Ward developed the idea (in his book *The Medea Hypothesis*) that the biosphere is not homeostatic, but rather in some ways self-destructive (Ward, 2009). Medea, like Gaia, is a mother figure from Greek mythology. But instead of symbolizing creativity and nurturing, Medea is famous for killing her own children. Ward's line of reasoning focused on the multiple extinction events in Earth's history that were not caused by collisions with asteroids.

The Great Oxygenation event and associated extinction event, about 2.5 billion years ago, is a prime example. It was induced by newly evolved bacteria capable of aerobic photosynthesis. The associated transformation of the atmosphere was toxic to most existing life forms. A second example is the end-Permian extinction event about 250 Mya (million years ago). It has been hypothesized (Kump, Pavlov, and Arthur, 2005) that the kill mechanism in this case was biologically based production of toxic hydrogen sulfide (derived from sulfate-reducing bacteria). Ward also emphasized the future fate of the biosphere. He cites global carbon-cycle model outputs that suggest biosphere-driven drawdown of atmospheric CO_2 will eventually prove fatal to the biosphere itself (Franck, Bounama, and von Bloh, 2006).

Lovelock built the Gaia hypothesis based on assumed negative feedbacks in the Earth system. For example, as solar luminosity increased over geological time, the biosphere compensated by reducing the concentration of greenhouse gases. Gaian homoeostasis relies in theory on cooperative interactions among biosphere components. In contrast, the Medea hypothesis looks to positive feedbacks, and to competition among biosphere components. In the Great Oxidation, microorganisms capable of aerobic photosynthesis and respiration outcompeted the primitive anaerobes and forced their retreat to the anaerobic ocean depths. Their exit left more resources (e.g., sunlight) for the aerobes.

Like the Gaia hypothesis, the Medea hypothesis has normative implications. Ward makes a connection between the dystopian tendencies of the biosphere and the current human-driven disruption of the Earth system. He pointed to *Homo sapiens* as a species that has found a winning strategy (use of fossil fuels) and is now laying waste to everything else in the biosphere, i.e., acting in typical Medean fashion. Ward's take-home message was that if we continue on this path (continuing to "rape and pillage the planet"), we could destroy ourselves. The conclusion is that we must change course and begin rationally managing the Earth system.

The Medea hypothesis did not receive nearly as much scientific or popular attention as the Gaia hypothesis. Ward never brought it to the peer-reviewed literature as Lovelock and Margulis had done with the Gaia hypothesis. Reviews of his book called it thought provoking, and indeed it does inspire one to consider the geological context of the human disruption of the climate system. However, the whole premise is rather weak because the biosphere has not destroyed itself over the 3.5 billion years of its life span, and, in fact, has consistently recovered from catastrophic disturbances and returned to high levels of biodiversity.

MECHANISMS OF BIOSPHERE INFLUENCE ON THE CLIMATE SYSTEM

Greenhouse Gases

As we noted in chapter 2, the terrestrial biosphere speeds up (by a factor of 10–100) the rock weathering process, hence pumping CO_2 out of the atmosphere. This amplification drives the carbon cycle to an approximate steady state at a lower atmospheric CO_2 concentration, hence a cooler mean temperature, than would be the case with the background purely physical-chemical system (Lenton, 2002). However, there is considerable disagreement about the magnitude of this cooling effect and how it has changed over geological time (Schwartzman, 1999). Presumably as the world gets warmer and wetter, this negative biosphere feedback strengthens.

How about at the cold end of the spectrum? When global climate is relatively cold, atmospheric CO_2 is generally low, and the biosphere in fact begins to be starved for CO_2 as photosynthesis decreases. As the vigor of the terrestrial biosphere declines, its capacity to enhance weathering weakens, and the CO_2 sink is moderated (Beerling et al., 2012; Pagani, Caldeira, Berner, and Beerling, 2009). Again, we have a negative feedback.

If we take a Medean view of the low CO_2 situation, the biosphere was largely responsible for driving CO_2 so low in the first place by speeding up the rate of mineral weathering. This systematic behavior shows up in the model of Franck and colleagues (2006), which Ward repeatedly cites. That model simulates how the interactions of the geosphere, biosphere, and climate regulate the carbon cycle over geological time and into the future. According to the simulations, the biosphere will face a CO_2 crisis approximately one billion years in the future because the area of readily weatherable rock will have increased and the biosphere will continue to speed up weathering rates. The biosphere will eventually become starved for CO_2 as well as being subject to very high temperature (driven by increasing solar radiation). Ward (2009) suggests that the biosphere will eventually drive CO_2 levels down so low that oxygen production will falter. Then the atmospheric O_2 level will decrease to the point of causing another extinction event, or indeed the demise of the biosphere. This scenario is a key argument for the Medea hypothesis.

Other mechanisms by which the terrestrial biosphere affects the atmospheric CO_2 concentration are through (1) increase or decrease in total biomass associated with changing climate and the area of ice-free land above sea level, and (2) formation of carbon compounds that are not readily decomposed and returned to the atmosphere (e.g., peat and coal). These processes operate at a range of spatial and temporal scales, which complicates attempts to understand the geological record of atmospheric CO_2 concentration.

Life in the ocean (the marine biosphere) also impacts the atmospheric CO_2 concentration. We've noted the ocean CO_2 sink associated with the biological pump, which occurs when marine organisms take up CO_2, and a portion settles to the ocean floor as organic carbon or calcium carbonate. Enhancement of this mechanism may have been part of a negative feedback to hyperthermal events such as the Paleocene–Eocene thermal maximum, mentioned in chapter 2. Faster weathering on land delivers more nutrients to the ocean, which increases marine productivity, which tends to draw down the atmospheric CO_2 concentration. Again, this is a characteristically Gaian mechanism.

In contrast to these negative feedbacks to climate warming, the approximately 80 ppm (parts per million) decrease in atmospheric CO_2 seen in the ice core record during the cold intervals of the glacial/interglacial cycle (see figure 2.4) is clearly a positive feedback. As discussed in chapter 2, it most likely reflects a transfer of CO_2 from the atmosphere to the ocean. A variety of processes would change the CO_2 in the ocean as the global climate warmed or cooled. As noted previously, cold water holds more CO_2; thus, simple diffusion tends to bring down the atmospheric CO_2 concentration as the climate cools. The ocean holds vastly more carbon that the land or atmosphere. Biologically driven CO_2 uptake in the ocean may also be enhanced. Hypothesized mechanisms for that process include increased delivery of iron and other nutrients to plankton in dust associated with the generally drier climate (Marinov, Follows, Gnanadesikan, Sarmiento, and Slater, 2008), or changes in ocean circulation that bring more nutrients to the surface. Greater ocean productivity means more dead organic matter settling to the ocean floor, with an associated increase in carbon sequestration. It thus appears that the marine biosphere participates in a positive feedback to climate change.

Methane and nitrous oxide (N_2O) are other greenhouse gases whose concentrations strongly depend on the biosphere. Although their concentrations are much lower than that of CO_2, these gases are respectively about 20 and 300 times more effective per molecule as greenhouse gases than is CO_2. They contribute 4–9 percent and 1–3 percent, respectively, to the current greenhouse effect. Microbes generate these molecules, and various chemical and biological processes break them down. By providing sources of these gases as a by-product of their energy metabolism, the associated microbes are providing a warming influence on the global climate.

The story is more complicated, of course, because other types of microbes are largely determining the oxygen concentration, and oxygen is involved in the chemical reactions that consume methane and nitrous oxide. The net effect of production and consumption is that their concentrations rise and fall in parallel with the global mean temperature during the glacial/interglacial cycles. Methane is produced by anaerobic respiration in wetlands, and tropical wetlands expand during warm interglacial periods. Likewise, for nitrous oxide. These greenhouse gases are evidently part of a positive biosphere-driven feedback to climate change.

The contribution of troposphere ozone, another significant greenhouse gas, is also complex. Earlier, we looked at stratospheric ozone because it plays a significant role in shielding the biosphere from ultraviolet radiation. However, ozone is also found in the troposphere (the lower atmosphere) at sufficient concentrations to contribute to greenhouse gas warming (it contributes 3–7 percent to the current greenhouse effect on Earth). The chemistry of tropospheric ozone is complicated, but one of the reactants involved in its production is a class of molecules called nonmethane hydrocarbons (NMHCs). These are low molecular weight gaseous compounds that volatilize from the surface of leaves and bark (Tingey, Turner, and Weber, 1991). The fragrance of a pine tree results from the release of NMHC molecules. Various physiological functions have been proposed for NMHC molecules, including chemical defense of plants, and a role in shunting excess solar energy off the main photosynthetic pathway under high light conditions. In any case, the concentration of ozone in the troposphere would be lower without biogenic NMHCs, and hence global climate would be cooler.

While researching this topic with funding from the U.S. Environmental Protection Agency's Global Change Research Program, I did an early assessment of possible feedbacks to climate change associated with NMHCs (D. P. Turner et al., 1991). Generally, higher temperatures mean greater NMHC production and more NMHCs mean more tropospheric ozone. Thus, the sign of the feedback to global warming would likely be positive. However, my study did not consider the recent ecophysiological studies that find high CO_2 may inhibit NMHC production (Beerling, Hewitt, Pyle, and Raven, 2007). The complexity of the coupling between the biosphere and the atmosphere in this case prevents us from specifying the sign and magnitude of the NMHC-mediated climate change feedback.

Surface Energy Balance

The biosphere also plays a poorly understood role in how clouds affect global climate. Clouds mostly reflect incoming solar radiation, thereby cooling the climate. The connection to the biosphere is in the provision of cloud condensation nuclei (CCN). Clouds form because water molecules condense into small droplets under the influence of decreasing air temperature. CCN are required to initiate cloud droplets, and on a non-biospheric Earth the CCN would most likely be dust particles. However, under some circumstances, dust particles can be in short supply, and this shortage may limit cloud formation. One illustration of this phenomenon is the observation of cloud trails behind oceangoing vessels. The air over the ocean usually is close to its saturation point (i.e., high relative humidity) and the provision of CCN from engine exhaust is sufficient to induce formation of clouds in the wake of a ship.

How does the biosphere produce CCN? One source is planktonic algae in the ocean that produce a molecule called dimethyl sulfide (DMS) as part of their osmoregulatory metabolism. When DMS molecules diffuse into the atmosphere, they interact with atmospheric oxygen to produce sulfate molecules. Aggregation of sulfate molecules results in aerosol particles that act as CCN. In fact, a large proportion of CCN over much of the ocean is derived from DMS. Thus, DMS has a significant overall cooling effect on global climate.

If there was a mechanism associated with global warming that favored DMS production, then biosphere-mediated negative feedback would occur (Charlson, Lovelock, Andreae, and Warren, 1987). However, recent modeling exercises do not support the case for strong global-scale negative feedback mediated by DMS, and the idea is still under debate (Levasseur, 2011). The observation that ocean acidification, associated with high anthropogenic CO_2, reduces DMS production (Six et al., 2013) does not bode well for a negative feedback mechanism to current CO_2-induced warming.

Over land, the shortage of CCN is not as acute as over the ocean, but biosphere production of CCN is important there, as well. The source of CCN in this case is the same NMHCs from plants that contribute to ozone production. In the atmosphere, NMHCs aggregate into aerosol particles that can act as CCN. They are often the dominant CCN over tropical rain forests (Chatfield, 1991). If global warming favored more tropical rain forests, there would likely be more clouds, and hence a cooling effect. Through this mechanism, biogenic NMHCs could thus participate in a negative feedback to global warming. To date, however, no one has worked out what the net sign of the NMHC feedback to global warming would be, using a method that considers both greenhouse gases and CNN effects. Efforts to understand the role of NMHCs in climate regulation exemplify the need for interdisciplinary research in Earth system science (Beerling et al., 2007).

The albedo (reflectance) of vegetation cover itself can influence climate significantly. Dark vegetation absorbs solar radiation, and the warming-driven conversion of tundra to conifer forests provides a positive feedback to climate warming (Beringer, Chapin, Thompson, and McGuire, 2005). In the temperate and tropical zones, the dissipation of absorbed solar energy by vegetation evapotranspiration provides a significant surface cooling effect.

There are also influences of vegetation on precipitation. A high rate of evapotranspiration by forests tends to keep the local atmosphere moist, and when the moisture condenses to form precipitation, this reaction "actively creates low pressure regions that draw in moist air from the oceans, thereby generating prevailing winds capable of carrying moisture and sustaining rainfall within continents" (Ellison et al., 2017, p. 53). Recycling of precipitation is estimated to support 20–35 percent of total precipitation in the Amazon River Basin (Eltahir and Bras, 1994).

BIOSPHERE INFLUENCE ON GLOBAL CLIMATE OVER GEOLOGICAL TIME

The Four-Billion-Year View

Both Lovelock and Ward point to the whole arc of biosphere history in support of their respective models. The pertinent facts are as follows: (1) solar irradiance has increased approximately 25 percent over Earth's nearly four-billion-year history, and (2) the concentrations of greenhouse gases have fallen by a factor of ten or more. The net effect has been a climate continuously able to support the biosphere.

Lovelock emphasizes that the biosphere has largely controlled the greenhouse gas concentrations in a way that compensated for the increase in solar radiation, and thus has acted in a homeostatic manner. At the time he proposed the Gaia hypothesis, Lovelock did not have a very clear idea about the specific mechanisms by which the biosphere could bring down greenhouse gases. But recent interest in biosphere-enhanced mineral weathering now fills that gap (Schwartzman, 1999). Lovelock's teleological Gaia seems to be most concerned about the planet overheating.

Ward's Medea hypothesis emphasizes the cold, low productivity threat to the biosphere. He suggests that the biosphere is practically on a death march, and will eventually (a billion years hence) kill itself by drawing down greenhouse gas concentrations too far.

The Gaian case for homeostasis is more solid on the face of it because it is trying to account for something that has already occurred. The Medean model is less compelling because it relies on projections a billion years into the future (Franck et al., 2006). Of course, neither view really explains how integration could be achieved across something as multifaceted as the biosphere.

The 500-Million-Year View

Considering only the past 500 million years, Earth system scientists have been able to assemble a reasonably coherent picture of the temporal variations in the geosphere, the biosphere, the atmosphere, and the climate. At the coarsest scale, there appears to be a rough oscillation between a warm state and a cold state. Under warm conditions ("Greenhouse Earth" or "Hothouse Earth"), CO_2 concentration is high (usually greater than 1,000 ppm) and glaciation is low. As a correlate of high CO_2, water vapor density would also be relatively high. The high greenhouse gas concentrations largely determine the global warmth. Conditions are reversed during the cold periods ("Icehouse Earth").

Two cycles of about 300 million years from Greenhouse Earth to Icehouse Earth are reasonably clear (figure 3.1). Some geologists also emphasize four cycles of approximately 150 million years. The most general explanation for the apparent cycles is based on plate tectonics. Fischer (1984) suggested a cycle between periods of mountain building and periods of quiescence. During mountain-building phases, volcanism is intensive and CO_2 levels are high (mostly greater than 1,000 ppm). Eventually large areas of unweathered minerals are exposed, and the associated high rates of mineral weathering (probably increased by the warm climate) draw down the atmospheric CO_2 concentration.

The most striking Icehouse phase is centered in the Permian period around 300 Mya. About that time, there is evidence in the geological record of massive glaciations and in the CO_2 record (e.g., from stomatal abundance) of a drop in CO_2 concentration (Retallack, 2002). Ward (2009) and other scientists point to a major change in the biosphere at that time,

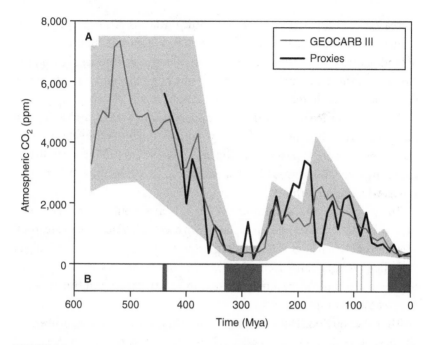

FIGURE 3.1

Comparison of estimated atmospheric CO_2 concentration and periods of glaciation throughout the Phanerozoic eon. Panel A has plots for the central tendency based on various proxies in the paleorecord, along with results and shaded uncertainty for GEOCARB III, a model that simulates the global carbon cycle over geological time. Panel B shows intervals of glacial (*dark shade*) or cool (*light shade*) climate. Mya = million years ago. Adapted from Royer, Berner, Montanez, Tabor, and Beerling (2004).

which likely accelerated the decrease in CO_2. Vegetation had been present on land since at least 500 Mya, but around 400 Mya, the fossil record indicates the evolution of plants with vascular systems and mycorrhizal (i.e., symbiotic with fungi) root systems. These developments had the effect of creating the pedosphere (soil), which greatly increased the surface area of minerals interacting with roots. The effect was an increase in the mineral weathering rate and a drawdown of atmospheric CO_2, i.e., a strengthening of the rock weathering thermostat (see box figure 2.1 in chapter 2). In addition, there is evidence in the geological record of increased coal formation (a carbon sink on land). This propensity to sequester carbon in coal can be attributed to the increasing productivity of land plants, and possibly the evolution of lignin synthesis. Lignin provided structural support for trees, but was not yet readily digestible or decomposable by heterotrophic organisms.

The Permian Icehouse phase was apparently not catastrophic for the biosphere. The climate did not get as cold as during a Snowball Earth state, there was not a mass extinction, and CO_2 probably did not fall below approximately 200 ppm. If Medea was going to kill itself, as Ward proposes, why not do it then?

The next cold phase, around 180 Mya, is also believed to have been driven by a CO_2 decrease associated with a high rate of silicate rock weathering. In this case, it was weathering of the supercontinent Pangea (Schaller, Wright, and Kent, 2015). The continent was moving north (driven by tectonic forces) into warm humid latitudes about this time, so the weathering rate sped up.

The other obvious Icehouse phase in the global mean temperature over the past 500 million years is relatively recent. It began about 30 Mya and continues to this day. In chapter 2, the various factors contributing to late Cenozoic cooling were discussed, in particular, thermal isolation of the poles, which favored glaciations, and the low CO_2 concentration (less than 500 ppm), which was probably brought down by vigorous weathering of the uplifted Himalayan Mountains (Ruddiman, 1997).

During the past million years, the climate has been about as cold as anytime in the past 500 million years (figure 3.1). The planet is apparently saved from going all the way to a Snowball condition by the strength of solar radiation (i.e., the Goldilocks effect) and negative feedback mechanisms such as low levels of water vapor and perhaps less cloud cover (associated with cold, dry air). Terrestrial biosphere enhancement of the rock weathering CO_2 sink is certainly diminished during the coldest periods (the glacial maxima) because the low CO_2 (approximately 200 ppm) reduces terrestrial biosphere vigor and the low temperatures reduce chemical reaction rates. The Milankovitch orbital forcings are evidently sufficient to initiate the glacial/interglacial cycles, but not strong enough to push the Earth system to Snowball Earth or Hothouse Earth states.

The four obvious Icehouse Earth phases do not support the hypothesis of strong biosphere control of global climate. Rather, tectonic processes that control CO_2 emissions (volcanoes, seafloor spread) and CO_2 sinks (the area of silicate rock available for weathering and its location) are dominant in these slow oscillations.

Climate crises such as the end-Permian extinction event and the Paleocene–Eocene thermal maximum may have more to say about the Gaia/Medea hypotheses. These periods appear to be induced by the geosphere, rather than a Medean biosphere. The termination of the hyperthermal episodes in the early Cenozoic period seems to have a significant Gaian aspect in that terrestrial biosphere–enhanced mineral weathering, and perhaps marine biosphere–enhanced carbon burial, both contributed to drawing down the atmospheric CO_2 concentration and returning the Earth system to cooler conditions. The same is true for the end-Permian extinction event.

The 100-Year View

In the past 100 years (the blink of the eye in geological time), the CO_2 concentration has increased from about 280 ppm to more than 400 ppm (a gain of 43 percent). The increase is undoubtedly driven by CO_2 emissions associated with fossil fuel combustion. This change in the atmosphere will have profound impacts on the biosphere, and from a Gaian perspective we would predict some sort of negative feedback, e.g., an increased net uptake of CO_2 to compensate for the increased emissions. As we shall see in chapter 4, this has largely not happened. Nor has the rate of silicate rock weathering significantly increased. Homeostatic Gaia has not shown up (Schlesinger, 2013).

If humans are considered part of the biosphere, then the ongoing upsurge in CO_2 concentration does appear to have a Medean character to it. As we shall see in chapter 5, the abruptness of the increase will take a toll on the biosphere by way of rapid climate change and ocean acidification. Note, however, that the biosphere is in no danger of annihilation. Much of its metabolism is microbial and will persevere even in the worst-case scenarios of human-induced environmental change.

Given the ambiguity of the historical relationship of the biosphere to the global environment over geological time, it is worth asking if we can identity any mechanism that might shape the biosphere into a superorganism-like entity capable of regulating Earth's climate.

BIOLOGICAL EVOLUTION AND BIOSPHERE EVOLUTION

One of the strongest scientific objections to Gaian claims of a homeostatic biosphere is that the concept does not fit well with contemporary interpretations of how biological evolution works (e.g., Dawkins, 1999). Traditionally, the unit of selection in biological evolution is the organism. In contrast, Lovelock's view points to the biosphere components of the Earth system as selected for how they contribute to the overall homeostatic capacity of the planet as a whole. Teleological Gaia would have to evolve a mechanism for coordination among its parts.

Lovelock and Margulis did not specifically address the evolutionary mechanisms by which Gaia was formed. They did point to cooperation within symbiotic relationships among species as an indicator that life can, in fact, be integrated above the organismic level. Margulis was the biologist on the Lovelock–Margulis team and she is best known as an early advocate for the theory that in the primitive biosphere, coevolution is likely to have been responsible for joining of multiple types of single-celled prokaryotes to form the larger, more complex eukaryotic type of cell (Margulis, 1970).

Formally, coevolution is a gene-for-gene interaction between two species; i.e., a mutation within the gene pool of one species favors or defends against a specific gene in the gene pool of another species. This kind of relationship is frequently found between pathogens and their hosts. The relationships can also be positive, as in the case of mutualisms in which both species benefit from the interaction. In the Margulis model of early microbial evolution, what had been free-living organisms became organelles within larger cells. However, these organelles kept their own genetic material (albeit with some losses). Selection thus began acting at the level of the genome—all genetic material within the larger organism.

The lichen is an often-cited example of mutualism. Here, two organisms are symbiotically linked, with the fungal and algal parts of the lichen dependent on each other for survival. (There are some exceptions; occasionally the algae can live independently.) Algal cells provide energy by photosynthesis and fungal cells serve both a structural function (providing a physical framework) and a metabolic function (promoting mineral weathering and nutrient uptake). This strategy has been solidly successful, and lichens have thrived for billions of years in a wide range of environments. Symbiosis between nitrogen-fixing microorganisms and plants verifies that nutrient exchange is a potent force in biological evolution.

Relationships between nonspecialized flowers and their pollinators provide examples of weaker coevolution. Many flowers depend on insect pollinators for their reproduction and they correspondingly invest considerable energy into attracting and rewarding those pollinators. Sometimes these weak relationships become quite strong. Famously, Darwin discovered a tubular flower with an extraordinarily long distance between the top of the flower and the base (containing the pollen and stigma). There were no know pollinators with a proboscis that long, so he predicted that such a creature must exist and would eventually be discovered. Indeed, many years later he was proved right!

One of the best-known examples of mutualism is the mycorrhizal association. In this case, a plant provides carbohydrates to a fungus that has infected its roots. The plant benefits because the fungal hyphae efficiently explore the soil for water and nutrients and transfer those resources to the plant. The intimacy of the interaction is such that in some cases the haustoria (root-like structures) of the fungus anatomically penetrate the fine root cells of the plant. Scientists have hypothesized that what started out as pathogenic relationship between species gradually evolved into a mutualistic relationship.

When microbes perform a nutrient cycling function, such as nitrogen mineralization (release of nitrogen from dead organic matter), they are ultimately providing nutrients to another ecosystem component that, in turn, nourishes them. Consider a soil arthropod (insect) that develops (through genetic mutation) an innovation allowing it to digest a previously indigestible component of tree litter. It uses the carbon-to-carbon bonds in the litter component as an energy source and excretes excess molecules (e.g., of nitrogen) for which it has no use from the litter. The tree life form then takes up the nitrogen and ultimately produces more litter that will eventually feed the soil arthropod.

This kind of nutrient cycling loop is the basis for the concept of an *ecosystem*. The term originated in the 1930s and is considered to include interacting biotic and abiotic components in a circumscribed location. Ecosystems require energy flow (usually a solar source) and are characterized by materials cycling. Among other pursuits, ecosystem ecologists develop nutrient budgets for the cycles of carbon, nitrogen, and other nutrients. As understanding of nutrient cycling improves, the sense of an ecosystem as a functional entity strengthens; e.g., ecosystem ecologists study mechanisms by which leakage of essential nutrients is accelerated after disturbances (Likens, Bormann, Pierce, and Reiners, 1978).

Ecosystems have been characterized as complex adaptive systems (Levin, 1998). In that view, the characteristics of the ecosystem (e.g., its trophic structure and nutrient conservation capabilities) are products of interactions among its components. Over time, these components become functionally linked by way of biological evolution because each is part of the selective environment of the others. The interactions of the components open the possibility of positive and negative feedback loops that reinforce the coupled system.

Studies involving artificial selection of miniature ecosystems hint at how ecosystems may change over time. In these experiments (Swenson, Wilson, and Elias, 2000), scientists constructed miniature ecosystems in small cup-like containers. To each container, a dollop of sterile soil was added that had been inoculated with a slurry of nonsterile soil (containing thousands of species of microbes). A fixed number of seeds of a simple plant were added. After the seeds germinated and were exposed to a month of sunlight, the experiment ended and aboveground biomass production—i.e., net primary production (NPP)—in each miniature ecosystem was determined by clipping and weighing. Soil lines that produced the most biomass were then combined (selected) to set up the next generation of containers, which were reseeded and allowed to grow again. After selection over multiple generations, some soil lines were consistently superior in the amount of NPP they could support. Note that the specific mechanisms of growth enhancement are not yet known; but most likely the mechanism involves enhancement of nutrient cycling by means of interactions among a specific set of microbes. In simulated microbial ecosystems, selection at the ecosystem level can be an effective adaptive force given a networked spatial structure (H. T. P. Williams and Lenton, 2008).

Experiments (intentional and otherwise) with introduction (Ehrenfeld, 2010) and removal (Ripple and Beschta, 2005) of species from more natural ecosystems also suggest strong interdependencies. In chapter 5 we will examine the phenomenon of trophic cascades, e.g., the process by which extirpation of wolves changes the behavior of grazers, which alters the composition and structure of associated forests. The emerging theory of niche construction is beginning to provide a basis for reconciling evolution and ecosystem dynamics. This concept (further discussed in chapter 4) refers to "organismal alteration of ecological patterns and processes in ways that confer heritable advantages and/or disadvantages to individuals or populations" (Ellis, 2015, p. 292). In other words, the operation of natural selection is based on both genes and environment impacts of those genes.

Ecologists traditionally think in terms of a levels of organization hierarchy (i.e., cells, tissues, organisms, populations, communities, and ecosystems). As we progress up that hierarchy, we leave behind the "myth of individuality" (Margulis, 2006), recognizing that higher organisms are indeed born as individuals, but if they are not highly integrated with other organisms in their vicinity, and with microorganisms within their own body, they may not last long. Part of the perhaps hyperbolic rejection of the Gaia concept by most evolutionary biologists was an ongoing tension within the discipline about the level at which natural selection operates (D. S. Wilson, 2001) and even what constitutes heritable variation and selection (Ellis, 2015). The controversy starts with disagreements about the "selfish gene" versus the organismic level (Noble, 2008), and extends much farther up the levels of organization hierarchy (Matthews et al., 2011). The Modern Synthesis for evolution accounts for Darwinian natural selection of organisms, Mendelian genetics, and population genetics. And later extensions accommodate the discovery of the biochemical basis for genetics in deoxyribonucleic acid. But there is now serious discussion of a fundamental extension of the Modern Synthesis paradigm for evolution; i.e., the new Extended Evolutionary Synthesis (Laland et al., 2015; Pigliucci, 2007) in which selection operates simultaneously at multiple levels of organization and evolutionary theory expands "beyond genetics to explain the evolution of complex phenotypic traits across a variety of taxa" (Ellis, 2015, p. 291).

Integration of Earth system science with evolutionary theory is only beginning to be thought about. Schwartzman (1999) refers to feedbacks between biotic evolution and biosphere evolution. Here's how it might work. Say a new biochemical pathway or trait, e.g., photosynthesis, is favored by natural selection. The photosynthesis trait is adaptive in its own right at the organismic level because it taps a new source of energy (i.e., compared to chemoautotrophs). The trait spreads throughout the biosphere such that the biosphere now begins to alter the global climate by sequestering CO_2, e.g., in marine sediments. The associated cooling of the global climate changes the selective regime, which may drive further biotic evolution toward life forms better adapted to the new climate. The evolution of the biota is driven by natural selection; the "evolution" of the biosphere is based on its functional role in the Earth system (box 3.1).

As we noted in discussing the early biosphere, no single way of making a living (i.e., type of metabolism) can persist for long because the available substrates become depleted, and rare nutrients tend to accumulate in living and dead organisms. Specialized organisms that can take advantage of the waste products of other organisms will be favored evolutionarily (Volk, 1998). The earlier noted evolution in the Paleozoic era of fungi capable of decomposing lignin, much of which at the time was turning into coal, is an interesting case study (Floudas et al., 2012; J. M. Robinson, 1990). Lenton and Watson (2011) refer to a series of biogeochemical revolutions in Earth history, generally associated with new forms of chemical recycling.

In Earth system science, we can consider Earth itself as the global ecosystem. The energy source is mostly the sun. Volk (1998) refers to biogeochemical cycling "guilds" as the living parts of the global ecosystem. These various types of organisms (e.g., nitrogen fixers) participate in the synthesis and breakdown of organic materials and maintain the global biogeochemical cycles. Thus, it is meaningful to talk about the

BOX 3.1
Semantic Issues

Earth system science is a transdisciplinary field, and as such is subject to occasional semantic issues when different disciplines use words in different ways. There is not necessarily a right and wrong here, but at times it is worth being especially alert to context.

From the perspective of studying the global biogeochemical cycles, the biosphere is considered the actual biomass of all life on Earth (Hutchinson, 1970). Its mass can be quantified, and it has a distinct functional role in maintaining the global biogeochemical cycles. For geographers, who may be more concerned with the spatial arrangement of their objects of study, the biosphere is considered the spherical space around Earth's surface that is occupied by life (Gillard, 1969).

An Earth system scientist speaks of the evolution of the biosphere, meaning changes in the array of species on Earth and their corresponding effects on the atmosphere and climate, e.g., the change from a biosphere dominated by anaerobic microorganisms to one dominated by aerobic microorganisms (Schwartzman, 1999). For an evolutionary biologist, the process of evolution refers to changes in gene frequencies within a population.

Ecologists speak of the biosphere as a complex adaptive system, meaning it is composed of coupled components—and the composition of the components as well as the nature of the couplings change in response to environmental change (Levin, 1998). For an evolutionary biologist, an adaptation is a genetically based trait that has been fixed in the gene pool by natural selection.

This issue of conceptual differences among disciplines plays out more broadly. Leemans (2016) refers to the difficulty in developing a conceptual framework for a transdisciplinary research program, "especially if it must be accepted by all different disciplines, who all, for example, differ in how they use central concepts, such as time. Physicists have adopted a continuous notion of time in their differential equations, ecologists and geographers use discrete time steps (years, seasons, generations) in their difference equations, while economists fold the future into the present using a discount rate" (p. 106). These sorts of differences can make transdisciplinary research trying at times, but certainly interesting.

"metabolism" of the biosphere, and about the biosphere as a component of the Earth system.

Although, there is only one biosphere and hence no possible analogue to Darwinian competition and selection at the global scale, it is worth noting that quasi-regulatory effects of the biosphere (or a particular species) on global climate might nevertheless be expected (Lenton, 1998, 2002). If the biosphere drove the global environment in a direction detrimental to itself (simply as a byproduct of its normal metabolism), then the biosphere as configured would lose vigor and hence be less effective in continuing to drive the environment in that deleterious direction. That seems to be the case when atmospheric CO_2 concentration falls below about 200 ppm (driven down in part by biosphere-enhanced weathering, which is a CO_2 sink). This type of negative feedback in the climate system relies only on the inherent property of living matter to grow under favorable conditions.

ESSENTIAL POINTS

The Gaia and Medea hypotheses are not truly scientific propositions. Neither makes sense from a traditional evolutionary perspective. However, both serve a purpose with respect to addressing global environmental change issues. The Gaia concept evokes the Earth system holistically—a planetary body that can be characterized in terms of energy flows and biogeochemical cycles. Conceivably, humanity can build a scientific understanding of the Earth system, which would provide a basis for assessing how it is being impacted by the technosphere and how a sustainable relationship of technosphere to biosphere might be constructed. The Medea concept also encourages a global-scale perspective. It reminds us that the Earth system as we know it is not permanent. Geosphere-induced and biosphere-induced changes can radically alter the chemistry of the atmosphere and oceans, as well as the climate. These changes are not always to the benefit of the dominant life forms.

IMPLICATIONS

1. The biosphere strongly influences the chemistry of the atmosphere, the oceans, and the soil; hence it participates in regulation of global climate. When geological or solar forces alter the climate, the response of the biosphere contributes to the net change in climate. The net biosphere response may be a negative or positive feedback. To understand and anticipate how the Earth system will respond to human-induced climate change, it is thus necessary to understand the response of the biosphere. Earth system science is pursuing that goal.

2. The biosphere strengthens the negative feedback built into the silicate rock weathering thermostat. However, the temporal framework of that feedback is on the order of thousands of years. Thus, it is not relevant in the near term to addressing the current pulse of climate warming associated with anthropogenic CO_2 emissions.

3. The influence of the biosphere on the climate is not always beneficial with respect to biosphere productivity and biodiversity. Humanity has come to expect the quasi-regulatory services performed by the biosphere, but they are not guaranteed as we shift the Earth system away from the state in which we inherited it.

4. Biological evolution changes the biota, in some cases introducing new biogeochemical cycling pathways. The spread of such changes throughout the biosphere can alter how the biosphere as a whole affects the chemistry of the atmosphere and the climate system, thus altering the selection regime for the biota. This integration of the evolution of the biota and the "evolution" of the biosphere is going on even now as *Homo sapiens*, a recent product of biological evolution, has begun (by way of the

technosphere) to radically alter the selection regime (e.g., by impacts on biodiversity and changing the global climate).

5. In the evolutionary history of the biosphere, "cooperation" has been as important as competition in driving the increase in complexity from bacteria, to eukaryotic cells, to multicellular organisms, to symbiotic associations among different species, and to ecosystems with their characteristic nutrient-cycling regimes. The cooperation metaphor points toward the needed integration of the technosphere with the rest of the Earth system to address major global environmental change issues.

FURTHER READING

Lenton, T., and Watson, A. (2011). *Revolutions that made Earth*. Oxford: Oxford University.

Lovelock, J. (1979). *Gaia, a new look at life on Earth*. Oxford: Oxford University Press

Ward, P. (2009). *The Medea hypothesis*. Princeton, NJ: Princeton University Press.

4

TECHNOSPHERE IMPACTS ON THE GLOBAL BIOGEOCHEMICAL CYCLES

The human expansion and domination of the Earth system over the past 10,000 years, albeit fleetingly recent in the 4.6 billion year context of Earth history, represents a major transition in the nature of life on Earth, comparable energetically to the colonization of land by plants.

—Y. Malhi (2014)

THE GLOBAL BIOGEOCHEMICAL CYCLES

Solar energy is used by the biosphere to generate and maintain order in the form of living organisms, ecosystems, and the biosphere itself. As noted in chapter 2, Earth is largely a closed system with respect to mass; thus, nutrients must be cycled to maintain biosphere productivity. Scientists speak of biogeochemical cycles because the chemical elements that make up living biomass are repeatedly transformed, moving in different chemical forms through phases in the geosphere and biosphere, as well as the atmosphere and hydrosphere in some cases.

The global biogeochemical cycles are each quite complex and involve myriad chemical compounds and chemical reactions. Many of those reactions are driven by the metabolism of a wide array of microbes and higher life forms, while others are purely inorganic. The biogeochemical cycles of some elements, such as carbon and nitrogen, are closely linked because both are assimilated by growing organisms. Despite the complexity of the global biogeochemical cycles, Earth system science has made considerable progress in characterizing the relevant chemical reactions, quantifying their rates at the local scale, and extrapolating (scaling up) the estimates for stocks (the amount present) and fluxes (transfers from one pool to another) to the global scale. This information is often summarized in box and arrow diagrams (e.g., figure 4.1).

The human enterprise (i.e., the technosphere) has added a new overlay of elemental transformations that tends to distort and to speed up the preexisting biogeochemical cycles (Zalasiewicz et al., 2016). Because the human overlay of biogeochemical transformations is a product of cultural evolution (box 4.1) rather than biological evolution, it can introduce

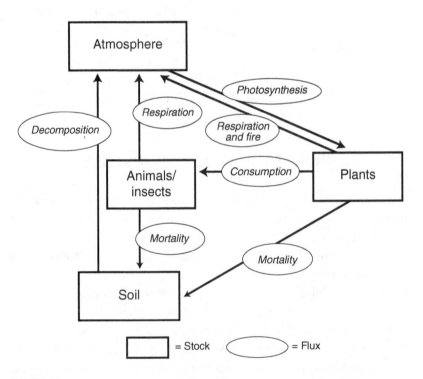

FIGURE 4.1

The fast carbon cycle on land. Stocks ("pools") that are "sources" supply carbon to another pool. "Sinks" receive carbon from another pool.

BOX 4.1
Cultural Niche Construction

How has one species, *Homo sapiens*, come to dominate the biosphere? The history of human influence on the biosphere and climate is well documented (e.g., Ruddiman, 2005), but the underlying mechanism is less clear. This question is worth asking since providing an answer in scientific terms could help us design a more sustainable relationship of humanity to the rest of the Earth system. Two key aspects of the answer are niche construction and cultural evolution (Ellis, 2015).

Niche construction was introduced in chapter 3 in the context of theoretical efforts to integrate ecosystem ecology and evolutionary biology. The ecosystem engineering of beaver is the paradigmatic example of niche construction. In that case, the way the beaver changes its environment affects the selection regime for itself and many other species. A young beaver is born into a constructed environment (its ecological inheritance).

Homo sapiens added a new wrinkle to reliance on niche construction by basing it on cultural evolution (learned behavior) rather than genetic evolution (mostly programmed behavior). The power of cultural evolution (i.e., based on the transmission of information across generations) is that it can proceed much more rapidly than genetic evolution. Like biological evolution, the products of cultural evolution are preserved across generations and build on each other (i.e., a "ratchet effect" prevails). With humans, the constructed niche (in terms of material culture such as buildings and ecosystem modifications such as agriculture) began to be specified by the products of cultural evolution. Ellis (2015, p. 293) refers to cultural niche construction as "alteration of ecological patterns and processes by organisms through socially learned behaviors that produce heritable advantages and/or disadvantages to individuals or populations."

> With humans, inheritance has thus come to include four components: genetic inheritance (e.g., big brains), cultural inheritance (e.g., information transmitted mostly by language), material inheritance (e.g., walled cities), and ecological inheritance (e.g., agricultural ecosystems). Thus, a baby is born into—and benefits from—a highly constructed environment, and is in a position to rapidly learn a tremendous amount about the world from the members of its community.
>
> This evolutionary strategy relies on language ability, and no other species has specialized in quite this way. A key problem now is that the ecological inheritance we are passing on has come to include characteristics of the global environment, and the traits by which our impact on the global environment are manifest may no longer be adaptive.

into managed and unmanaged ecosystems highly artificial concentrations and forms of chemical compounds, and very rapid rates of chemical reactions. Ecosystem-regulating mechanisms may thus become damaged, with consequent loss of system integrity. There is ample evidence for human-induced ecosystem dysfunction at the local scale (e.g., eutrophication of lakes), but we must now diagnose and treat possible global-scale disruptions.

The essential premise of Vernadsky's noösphere concept (introduced in chapter 1) is that humanity can learn enough about the global biogeochemical cycles to successfully manage the impact of the technosphere on them. Supporting an environment favorable to long-term human habitation of the planet depends on fulfilling Vernadsky's vision. Indeed, we must become a component in a teleological feedback to Earth system change (Lenton, 2016).

The text in this chapter is a synoptic survey, for the nonchemist, of human impacts on the biogeochemical cycles. It is intended primarily to give an indication of the degree to which the technosphere has altered their background operation. Smil (2002) and Schlesinger and Bernhart

(2013) go one step deeper in terms of explaining the chemistry and biology of specific elemental cycles and their global budgets. The text by Lenton and Watson (2011) discusses in greater chemical detail some of the interesting recent controversies in the global biogeochemistry literature.

THE EMERGENCE OF THE TECHNOSPHERE

The nutrient cycling capabilities of the biosphere have evolved over its approximately 3.7-billion-year lifetime. Much of the biochemical machinery that underlies the metabolism of microbes, plants, and animals is similar across life forms; e.g., the reliance on DNA for storing information, and on the set of 20 or so critical amino acids for building enzymatic compounds. However, biological evolution has vastly altered the structural forms and functional capabilities of distinct species.

One noted anatomical trend among animal species is toward ever more elaborate nervous systems and brains. *Homo sapiens* is clearly the leading edge of that trend. Nervous systems and brains offer many advantages in the Darwinian struggle for existence, and Professor of Evolutionary Paleobiology Simon Conway Morris nicely lays out the evolutionary sequence leading to the large brains in *Homo sapiens* (Conway Morris, 2015). Those brains are apparently required to support consciousness, and consciousness is required to generate a technosphere.

The emergence of "thinking matter" from the biosphere is as profound a change in the Earth system as was the emergence of living matter from the geosphere. We earlier noted the apparent order-friendly and life-friendly quality of the universe; from the astrobiology community, we also have the suggestion that we live in a consciousness-friendly universe (Martinez, 2014). Again, although an unscientific proposition (i.e., untestable), the implication is that we are in some sense "at home in the universe" (Kauffman, 1995) and conceivably have the potential to build a sustainable, high-technology planetary civilization.

It is worth briefly tackling the subject of consciousness here because this form of brain functioning is so important to understanding both the history of human interaction with the biosphere and our prospects

(Vandenburg, 1985). Consciousness is notoriously tricky to define, and to keep our level of abstraction manageable, let's mostly equate it with its highest form, i.e., self-awareness. One aspect of consciousness is the ability to know oneself as having existed in the past, as existing in the present, and as likely to exist in the future. Consciousness certainly involves thinking (cognition; i.e., the mental capacity to juggle words, concepts, and images). Consciousness implies the capacity for metacognition (thinking about or awareness of one's own thoughts). A related ability is the capacity for a "theory of mind": the ability to view another person as an intentional being and to imagine what he or she is thinking.

With consciousness, behavior can increasingly be shaped by social learning rather than genetic programming (which accounts for much of animal behavior). The earliest forms of thought, leading to a capacity for language, were based on imagination and mythology. Primitive language, along with a limited ability to associate cause and effect, was enough to transition from the hunter-gatherer way of life to one supported by agriculture. Gradually, thinking became more self-reflective and less referenced to anthropomorphic myths. Since about 1500 CE, we have gained the capacity for scientific thought, a key feature of which is the search for natural laws by way of observations and experimentation. The application of scientific thought transformed the relationship of humanity to the geosphere and biosphere. It allowed us to build the technosphere, a web of artifacts and machine-driven processes that has proliferated wildly over the surface of the planet in the past 100 years. As Vernadsky pointed out, we have become a geological force.

The emergence and ascendency of the technosphere is covered by the Building the Technosphere and Great Acceleration phases of the Anthropocene narrative (introduced in chapter 1). Building the Technosphere includes both the technological aspects of the Industrial Revolution and the socioeconomic changes associated with the rise of market capitalism. The Great Acceleration refers to the period after 1950 when multiple indicators of technosphere expansion and economic development entered a phase of exponential growth (figures 4.2 and 4.3; Steffen, Broadgate, Deutsch, Gaffney, and Ludwig, 2015a). Efforts to explain the extraordinary success of the human species in the context of the Extended Evolutionary Synthesis (introduced in chapter 3) are ongoing (box 4.1).

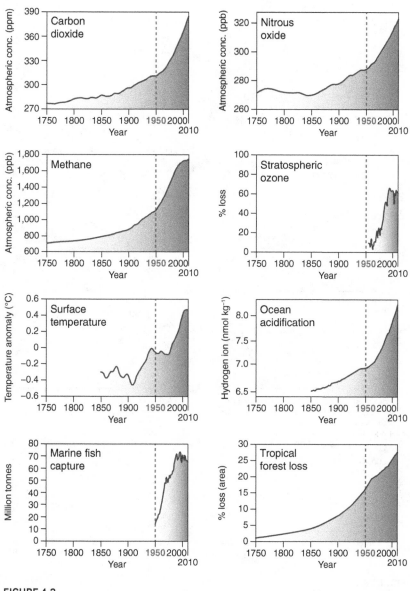

FIGURE 4.2

Earth system indicator trends (1750–2010) associated with the Great Acceleration.
Adapted from Steffen et al. (2015a).

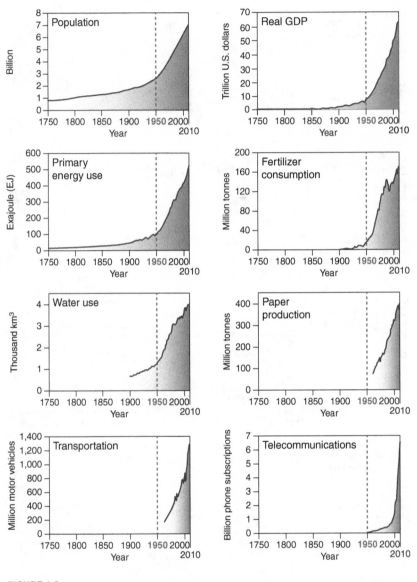

FIGURE 4.3

Socioeconomic indicator trends (1750–2010) associated with the Great Acceleration. Adapted from Steffen, Broadgate, Deutsch, Gaffney, and Ludwig (2015a).

CARBON

In chapter 2 we discussed the slow geochemical cycling of carbon (C) between a small pool in a gaseous form in the atmosphere, a large pool as a soluble form in the ocean, and the large below-ground pools in coal and oil formations, as well as in carbonate rocks. The carbon atom cycling time of this slow, mostly geochemical, pathway is millions of years. With the development of the biosphere, an alternative and much faster pathway of carbon cycling—among producers, consumers, decomposers, and the atmosphere—was established (figure 4.1). On land, carbon is taken up from the atmosphere by plants, stored in live biomass, transferred to a dead biomass pool at the time of plant death, digested by consumers or decomposer organisms, and eventually returned to the atmosphere. There is a parallel cycle in the ocean, with uptake dominated by photosynthetic plankton. The cycling time for this faster loop is years to centuries.

The technosphere has significantly disrupted both the slow and fast loops of the carbon cycle. The background global emissions of carbon from volcanism are on the order of 0.2 PgC (petagrams of carbon) per year (1 Pg = 10^{15} g, or 1,000,000,000,000,000 grams). Currently, the burning of fossil fuel plus industrial emissions (e.g., cement manufacture) introduces close to 9 PgC to the atmosphere each year (Le Quere et al., 2016); that is nearly 50 times the background geosphere source. The background carbon uptake (sink) associated with silicate rock weathering and the biological pump in the ocean (described in chapter 2) is presumably about the same magnitude as the tectonic source (0.2 PgC per year) because the atmospheric carbon dioxide (CO_2) concentration was mostly stable prior to the recent surge in emissions. The rate of silicate rock weathering is probably not increasing significantly in response to recent climate warming (Colbourn, Ridgwell, and Lenton, 2015).

With regard to the fast carbon cycle pathway, the technosphere is doing several things. First, we (its agents) are transferring large quantities of carbon directly from the live and dead biomass pools to the atmosphere in the process of deforestation. This transfer is on the order of 1.0 PgC per year in recent years (Le Quere et al., 2016). Areas of forest loss are mostly in the tropical zone, e.g., conversion to soybean fields in Brazil and to palm oil plantations in Indonesia. In the boreal zone, an increasing incidence of wildfire is also pushing up the carbon source from global forests (Kelly et al., 2013). The global soil organic matter pool is being reduced by agriculture; the soil organic carbon pool is typically reduced by half when native grassland vegetation is converted to agriculture (a function of both less input and faster decomposition associated with plowing).

Second, we are diverting 25–40 percent of the global terrestrial net primary production (NPP; the carbon fixed by plants and turned into biomass) from its previous pathway to local consumers (animals, insects) and decomposers, and using it for human purposes such as food, fuel, and fiber (Krausmann et al., 2013; Vitousek, Mooney, Lubchenco, and Melillo, 1997b). Global land NPP is about 60 Pg of C per year, only a factor of five larger than anthropogenic carbon emissions. This detour of NPP routes much of the fixed carbon directly back to the atmosphere rather than through the soil carbon pool (figure 4.1), thus tending to further diminish that pool.

One of the earliest discoveries of Earth system science regarding the global carbon cycle was that the terrestrial biosphere is now gaining carbon, despite deforestation and agricultural intensification. This conclusion is based on a simple mass balance calculation (table 4.1). Only three numbers are required. First, we must know the total anthropogenic carbon emissions. The fossil fuel and cement manufacturing component can be reliably estimated because fossil fuel is so economically important. Land use change emissions are more uncertain, but are relatively small. The second number is the amount of carbon accumulating in the atmosphere. It can be readily estimated by knowing the volume of the atmosphere and the annual increase in CO_2 concentration. Since the atmosphere is well mixed on annual time scales, its CO_2 concertation can be determined confidently with relatively few (about 100) measurement sites. Lastly, we must know the net uptake of carbon by the ocean. This is harder to estimate: it includes (1) the "solubility pump" (Raven and Falkowski, 1999), which is largely driven by the CO_2 concentration gradient between the atmosphere and the ocean; and (2) the "biological pump," previously discussed in chapter 3. Oceanographers are making many relevant observations and thus can reasonably estimate on an annual basis the ocean carbon sink.

TABLE 4.1 The Contemporary Anthropogenic Global Carbon Budget

FLUX TYPE	FLUX UNITS
Sources (emissions to atmosphere)	
Fossil fuel and cement	9.3
Land cover change/land use	1.0
Total	10.3
Sinks (uptake from atmosphere)	
Atmosphere	4.5
Ocean	2.6
Land (by difference)	3.1
Total	10.2

Note: Sources for these estimates and their uncertainties are covered in Le Quere et al. (2016). Flux units are petagrams of carbon (Pg C) per year. There is a 0.1 Pg C rounding error between total sources and total sinks.

Based on this mass balance approach, we are seeing a net uptake of carbon by the land in recent years (about 3.1 PgC per year). In the tropics, carbon accumulation in intact forests is compensating for losses from deforestation. In mid and high latitude forest ecosystems, forest inventories suggest that carbon is accumulating (Pan et al., 2011). However, the mechanisms of that accumulation are poorly understood (this is often referred to as the "missing carbon sink"). Some is from regrowth of forests associated with abandonment of marginal agricultural lands, as in the northeastern United States and in Russia. Wood is also accumulating in long-lived wood buildings. Other carbon sinks may be related to (1) enhancement of plant growth by the CO_2 itself (a key input to photosynthesis), (2) enhancement of plant growth by improved water use efficiency associated with high CO_2, (3) deposition of nitrogen (a common fertilizer and limiting plant nutrient) generated by the process of fossil fuel combustion, and (4) climate warming (Schimel et al., 2001).

Between the land and the ocean, only about a third to a half (depending on the year) of fossil fuel–generated carbon is accumulating in the atmosphere as a greenhouse gas (about 2.1 ppm [parts per million] per year). Remarkably, the proportion of emitted CO_2 each year that remains in the atmosphere has been stable on average over several decades (approximately 45 percent; Ballantyne, Alden, Miller, Tans, and White, 2012), which implies an increasing ocean plus land sink to compensate for the increasing anthropogenic emissions (11.2 PgC per year in 2015 compared with 1.5 PgC per year in 1950).

Both the land sink and the ocean sink are considered vulnerable. As the climate warms, we are seeing an increase in the area of forest burned each year in the boreal zone (van Lierop, Lindquist, Sathyapala, and Franceschini, 2015). These fires create both immediate CO_2 emissions and long-term emissions associated with decay of dead wood. Land carbon sinks may decrease from other factors, including ozone pollution, drought, and poor land management (e.g., overgrazing). The ocean sink (per unit of CO_2 increase) will decline as the ocean warms because of the simple physical fact that warm water holds less CO_2 that cold water. Climate warming may also impact ocean NPP and the biological pump by way of influence on ocean circulation and stratification (Steinacher et al., 2010).

The net effect of human influences on the carbon cycle has been a rapid increase in the atmospheric CO_2 concentration over the past 100 years (figure 4.2). The present concentration (2017) is approximately 410 ppm, compared with 260 ppm about 150 years ago as the Industrial Revolution accelerated, and 180 ppm about 15,000 years ago as Earth emerged from the last ice age (see figure 2.4). The current rate of increase in CO_2 concentration (approximately 2 ppm per year) is unprecedented in geological history (Zeebe, Ridgwell, and Zachos, 2016) and appears to be rising. The CO_2 concentration increase in 2015 was 3.05 ppm, the highest in the 56 years of measurements at Mauna Loa. As discussed in chapters 2 and 3, CO_2 is an important greenhouse gas and the effect of the increasing CO_2 concentration is an increasing strength of the atmosphere's greenhouse effect, thus altering the global climate (IPCC, 2014a).

Another problematic aspect of the contemporary carbon cycle is the trajectory toward ocean acidification (figure 4.4; Feely, Doney, and Cooley, 2009). As the CO_2 concentration in the atmosphere increases and more of it diffuses into the ocean, there is a shift in chemical equilibrium toward formation of a weak acid (carbonic acid), the product of interactions between CO_2 and water molecules. The problem is that as the ocean becomes more acidic, the chemical reactions by which coral reefs are built, and by which some free-floating planktonic organisms grow their calcium carbonate shells, do not work as well. The ecology of marine plankton will hence be altered, including patterns of NPP. Like the rate of CO_2 increase in the atmosphere, the rate of acidification of the surface ocean is likely unprecedented in geological history (because of the magnitude of the disequilibrium between the atmospheric and the whole ocean CO_2 concentrations).

Methane (CH_4) is next in importance after water vapor, CO_2, and ozone in maintaining Earth's background greenhouse effect (Kiehl and Trenberth, 1997). Its concentration has increased (figure 4.2) from 722 ppb (parts per billion) before 1750 to about 1834 ppb today, and it is

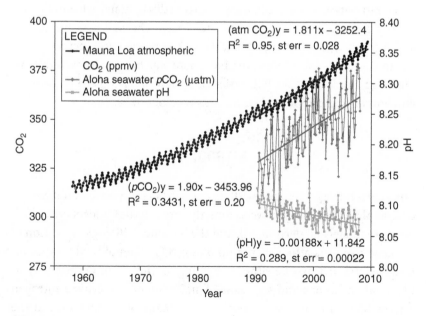

FIGURE 4.4

The recent history of atmospheric CO_2, ocean CO_2, and ocean pH. Modified with permission from Feely et al. (2009).

second only to CO_2 in contributing to the anthropogenic strengthening of the greenhouse effect (IPCC, 2014b). The cause of the increase in concentration is almost certainly anthropogenic, related to extensive rice cultivation (methanogens thrive in rice paddies), an increase in the population of domesticated ruminants that emit methane (e.g., cows), and leaks from the natural gas industry infrastructure. Natural sources are only about 40 percent of global methane emissions (GCP, 2017). Of special concern is the recent observation of increased methane emissions from the central United States (A. J. Turner et al., 2016). High latitudes are also receiving close attention because an increase there could signal the beginning of a significant positive feedback to global warming mediated by methane emissions from melting permafrost (Schuur et al., 2015).

As noted in chapter 1, Roger Revelle and Hans Suess characterized human alteration of the global carbon cycle as a large-scale geophysical experiment (Revelle and Suess, 1957). The recent dramatic increase in CO_2 concentration is beginning to look a lot like a geological event comparable to some of the greenhouse gas–driven disruptions earlier in Earth's history. For example, in the Paleocene–Eocene thermal maximum, CO_2 and methane emissions increased, the CO_2 concentration increased, the ocean acidified, and the marine biosphere was heavily impacted. How far we go along that path will be based on how long emissions continue and how strong carbon cycle feedbacks turn out to be.

NITROGEN

The global nitrogen (N) cycle differs from the carbon cycle in not having a mineral phase. Nitrogen cycles directly between the atmosphere (which is approximately 80 percent N_2), and the biosphere. Biological fixation of nitrogen from the atmosphere into biomass is accomplished by specialized bacteria that may be either free living (in the soil or ocean) or symbiotic (as in lichens and soybeans). Smaller amounts of fixed nitrogen are transferred from the atmosphere by means of precipitation to the biosphere during lightening events. Decomposition of organic matter generally releases a form of nitrogen available for uptake by other organisms, thus driving the ecosystem loop of producer, consumer, and decomposer.

Decomposition may also result in forms of nitrogen that are used by other specialized microbes as a substrate, and ultimately returned to the atmosphere.

Humanity has inserted itself into the global nitrogen cycle at several points (Vitousek et al., 1997a). In 1913, we began to fix nitrogen from the atmosphere into a plant-available form though an inorganic chemical process (the Haber–Bosch process). That chemical transformation requires a substantial input of energy, but the resulting fixed nitrogen has since been a boon to agriculture because crop production is often nitrogen limited. The background biological fixation of nitrogen at the global scale is approximately 170 TgN (teragrams of nitrogen) per year (1 Tg = 10^{12} g = 1,000,000,000,000 grams) and the input of plant-available nitrogen from lightening is much smaller (1–20 TgN per year). Through industrial processes we now fix approximately 100 TgN per year, and through cultivating crops such as soybeans we artificially fix about 25 TgN per year.

A second major pathway of human influence on the global nitrogen cycle is release of fixed nitrogen (i.e., not N_2) to the atmosphere by combustion of fossil fuels. The magnitude of this flux is on the order of 25 TgN per year. The form of nitrogen produced by fossil fuel combustion is a molecule with one nitrogen atom and one or more oxygen atoms (NOx). Nitrate (an ionic form with three oxygen atoms) is very soluble (i.e., mixes well with water), and tends to be returned to the surface in precipitation as a dilute form of nitric acid. This form of nitrogen represents an artificial fertilization of unmanaged ecosystems and contributes to soil and aquatic acidification. Biogeochemists have developed the concept of nitrogen saturation at the ecosystem scale, referring to the level of artificial nitrogen deposition at which increased plant uptake cannot keep up, and deleterious effects begin to predominate (Aber et al., 1998).

In the United States and Europe, poorly regulated fossil fuel combustion in the mid-twentieth century led to high enough deposition of nitric and sulfuric acid (similarly produced) in precipitation to become a political issue. National and international agreements to scrub nitrogen and sulfur emissions from smokestacks and vehicles helped moderate the impacts on soils and water bodies (see chapter 8). Nevertheless, soil and aquatic acidification remain a significant environmental issue in the United States and Europe. This is even more of a concern in China and

India, where fossil fuel combustion is still growing and associated nitrogen emissions are less regulated.

Combining nitrogen fixation from the fertilization industry, cultivation of nitrogen-fixing plants, and emissions from fossil fuel combustion yields a total of about 150 TgN per year, an amount of comparable magnitude to the background biological fixation. As noted earlier, one hypothesis for why the biosphere is currently sequestering carbon in some regions is that this massive subsidy of anthropogenic nitrogen is acting as a fertilizer in unmanaged ecosystems.

Another consequence of human intervention in the global nitrogen cycle is manifest by way of greenhouse gases. The nitrate form of nitrogen, besides being taken up by plants, can also be used by specialized microbes in their energy metabolism and transformed to nitrous oxide (N_2O). This chemical form of nitrogen is a potent greenhouse gas, and its concentration in the atmosphere is increasing in recent years (figure 4.2), contributing 6 percent to current climate change (Lashof and Ahuja, 1990). Human activities contribute about 40 percent to the current global total annual N_2O source. Nitrate-consuming microorganisms are most likely to capture nitrate when it is present in amounts beyond what plants need. Thus, the increase in N_2O is very likely driven in part by excessive agricultural nitrogen fertilization.

Nitrogen oxides released by fossil fuel combustion also contribute to the formation of ozone in the lower atmosphere (the troposphere). Photochemical smog in large metropolitan areas is characterized by high ozone levels, and in sufficient concentration it is toxic to plants and animals. Like methane and nitrous oxide, ozone is a significant greenhouse gas and its increasing concentration in the lower atmosphere (the troposphere) is contributing to the total greenhouse gas–induced global warming.

PHOSPHORUS, SULFUR, AND CALCIUM

The global cycle of phosphorus (P) differs from the global cycles of carbon and nitrogen in not having a gaseous phase. Phosphorus becomes available to the terrestrial biosphere through chemical weathering of bedrock minerals. Humans currently mine and grind up rocks of high phosphorus content to produce phosphorus fertilizers for cropland.

The human-driven input of phosphorus to the biosphere is about triple the background delivery from mineral weathering (Smil, 2002). Natural phosphorus is not very soluble, so it tends to remain in place after being weathered out of minerals. It can be transported in streams and rivers to the ocean, where it is cycled by the ocean biota before sequestration in ocean sediments and incorporation into a mineral form.

Fertilizer phosphorus is delivered in a mobile form to facilitate uptake by crop plants. When supply and demand are not in synchrony or of the same magnitude, phosphorus tends to leach out of soil and, like nitrogen, contributes to eutrophication (excess productivity leading to oxygen depletion) in rivers and lakes. Deposition of phosphorus-rich human and animal wastes into water bodies exacerbates this problem. The high levels of phosphorus and nitrogen in large rivers that drain agricultural regions create so-called dead zones in coastal waters in some cases (Diaz and Rosenberg, 2008). At the mouth of the Mississippi River, a large dead zone forms each year for several months. During that period, the level of oxygen throughout the water column is too low to support fish and crustaceans.

The story of the global sulfur (S) cycle is similar in some respects to the nitrogen cycle in that the release of sulfur through combustion of fossil fuel (about 80 TgS per year) is large compared to background fluxes. Volcanic emissions to the atmosphere are only 14 TgS per year (Smil, 2002). Like nitrate, sulfate in the atmosphere from fossil fuel combustion precipitates out rapidly in an acidic form.

Sulfate molecules tend to aggregate and form aerosol particles in the atmosphere. These particles reflect solar radiation and become cloud condensation nuclei, and thus can have a significant cooling effect on the regional and global climate. The eruption of Mount Pinatubo in 1991 ejected enough sulfate into the atmosphere to cool global mean temperature 0.2°C–0.5°C for several years. Sulfate aerosols also influence ozone chemistry in the atmosphere and contribute to depletion of stratospheric ozone (Robock, 2000). Anthropogenic sulfate inputs to the atmosphere are currently having a significant cooling effect on global climate. However, the magnitude of the cooling effect is difficult to estimates because of the extreme spatial heterogeneity of the industrial emissions and the short atmospheric lifetime of sulfate—which is too short to be well-mixed throughout the earth's atmosphere (Wilcox, Highwood, and Dunstone, 2013).

Other plant nutrients such as calcium, magnesium, and potassium are, like phosphorus, derived from rocks that are ground up for use in fertilizer. Also like phosphorus, they tend to be applied as fertilizer in a mobile form, and thus are susceptible to leaching. When irrigation is used improperly, these minerals (in fertilizer or brought to the surface with evaporative flow) are left as salts at the surface during periods of rapid soil evaporation. This salinization is a significant form of soil degradation. Poorly managed irrigation and associated salinization has made approximately 10 percent of globally irrigated land unusable for agricultural purposes, and the area of degraded land is increasing (McNeill, 2000).

METALS

The rapid evolution of the technosphere has depended on our ability to isolate and manipulate elements such as gold, copper, iron, and aluminum. These elements were isolated from rocks, purified, and combined in alloys in ever-increasing amounts over the course of human history. Examination of the global budgets for many of these metals suggests that humans mobilize as much or more than the background fluxes (Rauch and Pacyna, 2009).

Of particular interest with respect to the biosphere are those elements that are toxic at low concentrations. Lead began to be added to gasoline beginning in the 1920s to increase automobile engine performance and life. However, based on accumulating evidence of its impacts on human health (neurotoxicity), its use as an additive was halted, resulting in a large decrease in anthropogenic emissions to the atmosphere. A decline to virtually preindustrial concentrations shows up nicely in ice cores from the Greenland ice sheet (Boutron, 1995).

Mercury (Hg) is also toxic and is added to the atmosphere by way of fossil fuel combustion (mostly coal) and to water bodies by way of industrial applications (mining). Global anthropogenic mercury emissions are at least an order of magnitude greater than natural emissions (Nriagu, 1989) and only 17 percent of the mercury in the contemporary ocean is of natural origin (Amos, Jacob, Streets, and Sunderland, 2013). Because of its tendency to bioaccumulate, mercury can reach toxic concentrations in freshwater and marine fish. As with lead, there has been enough research concerning its health effects on humans and wildlife to justify tighter regulation of mercury emissions from power plants (in some countries).

Other trace metals, or heavy metals, that are potentially toxic and are now being released into the environment in increasing quantities include cadmium and chromium. They are extracted from rocks for industrial purposes. Recently there has been increasing concern in China with movement of cadmium from mining wastes into freshwater, and then into the human food chain by way of irrigated crops (Hsu and Miao, 2014).

RADIOACTIVE ELEMENTS

Radioactive atoms spontaneously emit high-energy electromagnetic radiation and particles (e.g., electrons) during decay from one element form to another. These atoms originated in nuclear fusion reactions, mostly in stars. Radioactive uranium was concentrated in Earth's core as the planet solidified, and energy from its decay and that of other radioactive materials drives the tectonic cycles discussed in chapter 2.

Scientists first discovered radioactivity in the context of X-rays in the late nineteenth century. Physicists gradually worked out the processes by which radioactive atoms were formed and decay. Besides fusion reactions in stars, they could form in the atmosphere by way of interactions of atoms with high-energy solar radiation or cosmic radiation (from outside the solar system). By World War II, physicists realized the potential for explosive chain reactions if radioactive atoms were concentrated. The successful construction of atomic bombs helped end World War II, but also set off a nuclear arms race associated with the subsequent Cold War.

Aboveground nuclear testing in the 1940s and 1950s pushed up the atmospheric burden of a variety of short-lived and long-lived radioactive species. Experiments with animals had shown that inhaled or ingested radioactive atoms can induce various forms of cancer. The realization that aboveground nuclear testing was a significant health threat prompted an international agreement to curtail it (see chapter 8). The atmospheric concentrations of bomb-generated radioactive materials have declined, but the long-lived chemical species originating in bomb testing can still be detected in soil.

The nuclear power industry has also been a source of radioactive species to the environment. Europe became aware of the nuclear power plant accident in Chernobyl when radioactivity detectors began lighting up as the radioactive fallout blew in from the east. The sites around nuclear power plant accidents such as Chernobyl and Fukushima are heavily contaminated with radioactive material and must be closed to human habitation.

OXYGEN

Earth's atmosphere is 20.8 percent oxygen (O_2). We've noted that James Lovelock formulated the Gaia hypothesis in part because there was so much oxygen in the atmosphere despite it being quite chemically reactive. The source of most oxygen in the atmosphere is the splitting of water molecules during photosynthesis. The primary sink of oxygen is metabolic respiration, in which oxygen reacts with energy-rich organic compounds (e.g., sugars). Water is a by-product of aerobic respiration; hence oxygen cycles between the atmosphere, the biosphere, and hydrosphere. An alternate oxygen sink is combustion of biomass (i.e., fire).

By combustion of fossil fuels, the technosphere has boosted the consumption of oxygen and correspondingly the production of CO_2 (hence global warming). However, since the concentration of oxygen is orders of magnitude greater than that of CO_2, there has been minor impact on the oxygen concentration in the atmosphere. Ralph Keeling, son of David Keeling (who first detected the increase in atmospheric CO_2) devised an oxygen-monitoring instrument sensitive enough to detect the slight decreasing oxygen trend in recent decades that corresponded to the CO_2 increase (and an annual cycle linked to the annual oscillation of the CO_2 concentration). Earth system scientists have no concerns about oxygen depletion from fossil fuel combustion, or possible reduction in oxygen production because of human impacts on global photosynthesis. There is, however, concern about declining oxygen in the ocean, associated with climate warming, and possible biological impacts (R. F. Keeling, Kortzinger, and Gruber, 2010).

Atmospheric chemists also have concerns about two more subtle aspects of oxygen chemistry. The more familiar issue is stratospheric ozone depletion (Rowland, 2006). The subtler issue involves hydroxyl radical depletion (IPCC, 2007). Both issues center on the extraordinarily complicated chemistry of oxygen in the atmosphere, one of the more formidable topics in Earth system science.

We have noted the role of ozone (O_3) as a greenhouse gas but it is also important to the Earth system because in the stratosphere it absorbs ultraviolet (UV) wavelengths of solar radiation, in effect protecting the biosphere from this highly energetic form of sunlight. A "hole" in the stratospheric ozone layer was discovered in the 1980s when scientists concluded that instruments in Antarctica that had been monitoring the column abundance of ozone and recording anomalous drops during the winter since the 1970s were not faulty, but were registering accurate numbers after all. Indeed, the level of ozone in the stratosphere during the polar winter was much lower than had been previously measured. Atmospheric scientists had already identified a possible mechanism of ozone depletion involving chemical reactions driven by anthropogenically created chlorofluorocarbons (CFCs). These compounds were developed in the 1930s for use as nontoxic, nonflammable refrigerants and propellants. Their inert nature made them ideal for those purposes, but an unintended consequence was that they accumulated in the atmosphere. After slow diffusion to the high-energy environment of the stratosphere, they broke down, and the chlorine became chemically reactive as a catalyst to ozone consumption. Here, clearly, the technosphere was threatening the global life support system.

It took the atmospheric chemistry community a decade or so to understand what was causing stratospheric ozone depletion, then more years to convince the policy community to do something about it. But with the ratification of the Montreal Protocol (discussed in chapter 8), the industrial nations of the world agreed to stop CFC production. The treaty has largely been effective, and the CFC concentration is slowly decreasing (figure 4.5), with an associated shrinking of the ozone hole. The only downside to the story is that the replacement refrigerants are also long-lived compounds that accumulate in the atmosphere, and they are strong greenhouse gases.

The hydroxyl radical (OH) is the so-called cleanser of the atmosphere. It is a very short-lived product of photochemical reactions involving oxygen. OH attacks a wide variety of reactive carbon and nitrogen molecules and converts them to forms that leave the atmosphere through precipitation. The potential problem is that the technosphere is loading up the atmosphere with so many reactants (pollutants), that the availability of OH may be decreasing. Over half the global emissions of reactive carbon and nitrogen compounds are now anthropogenic in origin (Lelieveld, Dentener, Peters, and Krol, 2004). The consequence of high anthropogenic emissions of pollutants is that the atmospheric lifetime of molecules such as methane (a greenhouse gas) is extended. The complexity of the atmospheric chemistry and the extremely brief lifetime of the OH radical molecule keep this an active area of Earth system science research.

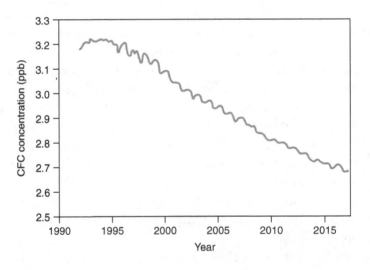

FIGURE 4.5

Time series for effective equivalent chlorine concentration (a metric related to capacity to deplete stratospheric ozone); ppb = parts per billion. Adapted from NOAA (http://www.esrl.noaa.gov/gmd/hats/graphs/graphs.html).

THE HYDROLOGIC CYCLE

Humans are master manipulators of water. We now use in one way or another approximately 50 percent of the global freshwater flow in streams and rivers (Goodie, 2006; McNeill, 2000). Some of that water is returned to the river from which it was appropriated (e.g., downstream from large cities), but often in a polluted form.

Large dams and associated irrigation are now found on almost all major river systems on the planet and have the effect of increasing evaporation, which reduces downstream river flow. The Aral Sea in Central Asia has virtually disappeared as of result of diverted inflows. Large river deltas all over the planet are starved of silt by upriver dams, and subject to salt water intrusion by reduced flows.

Another indirect human influence on stream flow is a shift in the pattern in spring runoff. As the climate warms (from increasing greenhouse gas emissions) in mountainous area, more precipitation falls as rain rather than snow, consequently ending up in stream flow earlier (Barnett et al., 2008). What snow does fall also tends to melt earlier, thus providing less of a sustained stream flow during the growing season. Glaciers are in retreat globally (NSIDC, 2013) and it is anticipated that their demise will have significant impacts on seasonal availability of water in several heavily populated regions, such as the Ganges River Basin in India.

In some regions, the quantity of water extracted from rivers is augmented by significant amounts from underground aquifers. Depending on depth, belowground aquifers are naturally recharged, but often at nothing like the rates at which water is now being extracted. The deepest aquifers are essentially fossil water, and these are largely being mined rather than used sustainably (Pearce, 2006).

The story of the Ogallala Aquifer in the United States is a great case study. This underground reservoir underlies much of Nebraska and stretches south to western Kansas, Oklahoma, and Texas. Extraction is largely unregulated, and wells feeding center point irrigation agriculture have proliferated in recent decades. The aquifer is being rapidly emptied in some areas; consequently, withdrawals have been curtailed by the requirement to dig ever deeper wells. A similar story is unfolding in the Central Valley of California and in the Indus River Basin in India.

Remarkably, there is now a satellite-borne sensor (GRACE, further discussed in chapter 9) that can monitor the size of groundwater reservoirs based on faint signals in the gravitational field (Strassberg, Scanlon, and Chambers, 2009).

Besides water on the surface and in the ground, we are influencing the amount of water in the atmosphere by extensive changes in land cover. Deforestation reduces the rate of water return to the atmosphere via transpiration and changes the local energy balance. In Costa Rica, deforestation over large areas in the eastern coastal plain is resulting in higher clouds and a drying of the downwind mountain cloud forests (U. S. Nair, Lawton, Welch, and Pielke, 2003). In the Amazon Basin, regional climate model simulations suggest that continued deforestation will set up a positive feedback loop such that large areas of deforested land will reduce local precipitation and cause degradation of adjacent intact rain forests (Costa and Foley, 2000).

At the global scale, the warming of the atmosphere associated with human-induced rise in the concentrations of various greenhouse gases is increasing the amount of water vapor (another greenhouse gas) held in the atmosphere. Since warm air holds more water than cooler air, and we live on a planet whose surface is two thirds water, there is increased evaporation into the atmosphere over water as climate warms (Chung, Soden, Sohn, and Shi, 2014). This positive feedback to climate warming was noted in chapter 3.

More water vapor leads to an associated increase in global precipitation. However, that overall increase is playing out as big increases in some areas and decreases in other areas. One pattern that is both predicted by the general circulation climate models that simulate the effects of rising greenhouse gas concentration, and is beginning to be observed by networks of meteorological stations, is an ongoing drying of the lower mid latitudes (e.g., Neelin, Munnich, Su, Meyerson, and Holloway, 2006; Seager et al., 2007). This drying trend may already be having significant impacts on vegetation (Breshears et al., 2005).

A quite pervasive but poorly understood indirect human influence on the hydrologic cycle is the increasing water use efficiency of vegetation (Keenan et al., 2013). In chapter 2, we noted that one of the paleo-proxies for atmospheric CO_2 concentration is stomatal density in fossilized leaves.

Plants tend to reduce stomatal density as CO_2 concentration increases because diffusion of CO_2 through the stomata is faster. A key benefit of low stomatal density is less evaporation of water vapor out of the leaf (i.e., transpiration), and therefore less sensitivity to drought. The effect on transpiration at the canopy scale is difficult to predict, however, because vegetation may also increase leaf area (Tng et al., 2012). The effect of reduced transpiration at the watershed scale is local warming, because solar radiation is absorbed and reradiated rather than being used for transpiration. Less transpiration could also potentially result in an increase in stream flow (A. J. Turner et al., 2016; Wiltshire et al., 2013).

Lastly, let's consider changes in the ocean. The approximately 1°C increase in global mean temperature over the past century has already significantly influenced sea level. Satellite monitoring of glaciers and ice sheets has documented a global decline in volume that explains part of the increase. Thermal expansion as the ocean warms explains the rest. The current rate of sea level rise is approximately 3 mm (millimeters) per year, and the rate is accelerating (IPCC, 2014b, Watson, 2016). The total increase since 1901 is approximately 200 mm (close to 8 inches). The impacts of sea level rise on the technosphere are limited at this point, but there is every reason to believe that we are at the very beginning of a rise in sea level that will amount to many meters and play out over hundreds of years (see chapter 6).

We are also altering the ocean's thermohaline circulation (THC). This so-called "ocean conveyor belt" brings warmth to high latitudes in the North Atlantic. The underlying mechanism is based on flow of the Gulf Stream current along the east coast of North America, which is partially driven by the Atlantic Ocean component of the THC (the Atlantic meridional overturning circulation). This flow keeps Europe warmer than would be the case without it (Palter, 2015). Typically, as water in the Gulf Stream cools at high latitude, it becomes denser and sinks, which maintains the flow of the THC (Broecker, 1987). However, with global warming–induced melting of the Greenland ice cap, and possibly more precipitation because of the intensification of the hydrologic cycle (Huntington, 2006), more freshwater is input to the North Atlantic, causing a freshening of those waters, and hence a decrease in density and less propensity to sink (Rahmstorf et al., 2015). In addition, the disproportionate heating of the polar regions by greenhouse gas warming reduces the north-south temperature gradient between the tropics and the high latitudes, which diminishes the Gulf Stream. The impacts of a slowed THC include a cooler Europe and sea level rise along the east coast of North America.

ESSENTIAL POINTS

The technosphere juggernaut is rapidly altering many of the global biogeochemical cycles. Vernadsky was certainly prescient in recognizing as early as the 1920s that humanity is the equivalent of a new geological force. Since the 1950s, especially, there has been rapid acceleration of mass fluxes through the technosphere and into the atmosphere, biosphere, and hydrosphere. Besides altering the flux rates of many background biogeochemical cycling reactions, the technosphere has brought to Earth's surface (our biosphere) high levels of toxic elements such as cadmium, and broadly distributed radioactive elements such as cesium and uranium. Environmental problems arise when the background mechanisms of biogeochemical recycling are overwhelmed (or do not exist) and undesirable waste products accumulate. Technosphere impacts include a rapid increase in the atmospheric concentrations of the most important greenhouse gases, and intensification of the hydrologic cycle.

IMPLICATIONS

1. The reflexive consciousness of *Homo sapiens* (thinking matter, if you will) is a qualitatively new addition to the several billion-year-old biosphere (living matter). The capacity of humans to work collectively in large social groups, and to think in a scientific manner, has led to a massive disruption of the global biogeochemical cycles. The same brainpower and technological prowess that has induced a crisis in natural resource management must now devise a more sustainable relationship of technosphere to biosphere (the focus of chapters 7–10).

2. The growth of the technosphere is inducing a warming of the global climate. For now, the reciprocal impacts of that climate change on the

technosphere have been minor. However, at some point we may see a negative feedback loop emerge such that global environmental change begins to inhibit the growth of the technosphere.

3. The current geologically unprecedented rate of increase in atmospheric CO_2 is inducing rapid acidification of ocean surface waters. This change in ocean chemistry is beginning to alter ocean productivity and the ocean carbon sink. Thus, there are two compelling reasons (climate change and ocean acidification) to rapidly reduce fossil fuel carbon emissions.

4. Sulfur emissions associated with fossil fuel combustion produce sulfate aerosols with a significant cooling influence on the current climate. When burning of fossil fuels is constrained, some additional warming will be manifest because of the reduction in sulfate aerosols. This additional warming must be planned for.

FURTHER READING

Archer, D. (2010). *The global carbon cycle*. Princeton, NJ: Princeton University Press.
McNeill, J. R. (2000). *Something new under the sun*. New York: Norton.

5

TECHNOSPHERE IMPACTS ON THE BIOSPHERE

*The world is no longer characterized by "natural ecosystems"...
with humans disturbing them, but rather is characterized by a
combination of human-engineered and used ecosystems and
more or less modified novel ecosystems.*

—M. Williams et al. (2015)

Humans have an innate tendency to live in social groups and alter the local environment. This propensity for living in groups is not unique among animal species. In the tropics, termites build mounds up to 7 meters tall housing thousands of "citizens." Leafcutter ants likewise build vast underground chambers for cultivation of the mushrooms that provide most of their sustenance. They harvest significant amounts of vegetation near their nests and alter the local landscape. Behavior of these insects is largely genetically programmed, and their way of life does not change much from generation to generation (E. O. Wilson and Hölldobler, 2008). The unusual feature of human behavior is that we consciously try new behaviors, and if successful, transmit them horizontally (within a generation) and vertically (across generations) by way of social learning.

That mental flexibility and capacity for transmitting information has given humans the ability to colonize almost the complete range of climatic zones on Earth. Even as hunter-gatherers, we adapted to everything from

Arctic tundra to tropical rain forests. With the development of the technosphere by way of cultural evolution, we have not retreated from anywhere, and our population density has increased everywhere. Thus, the scale of our impacts on the environment has extended from the local to the global level.

Efforts to assess human-mediated impacts on the biosphere, such as the United Nations–sponsored Millennium Ecosystem Assessment (MEA, 2005) and the Intergovernmental Science-Policy Platform on Biodiversity and Ecosystem Services (IPBES, 2017), have attempted to integrate social science along with indigenous or traditional knowledge to a greater degree than efforts to assess human impacts on climate (e.g., the Intergovernmental Panel on Climate Change [IPCC]). This more transdisciplinary approach creates a tension about what constitutes knowledge, and how best to approach adaptation to global environmental change (Obermeister, 2017). The globally valid knowledge base of biophysical science is increasingly augmented with the local knowledge involving specific people and places (Escobar, 2001).

LAND COVER AND LAND USE CHANGE

Human impacts on the land surface include changes in land cover type, e.g., from forestland to agricultural land, as well as changes in land use, e.g., primary (native) forestland to plantation forestland (Lambin, Geist, and Rindfuss, 2006). As discussed in chapter 4, these changes influence the local hydrologic cycle (DeFries and Eshleman, 2004), regional climate (Pielke et al., 2011), and the global carbon cycle (Houghton et al., 2012). They also determine the geographic distribution of many wild and domesticated plant and animal species.

The importance of human-induced land surface change has inspired sustained attention from the Earth system science research community (B. L. Turner et al., 1993), notably by the Land Use and Land Cover Change Project of the International Geosphere-Biosphere Programme (Lambin et al., 2006). Ambiguities in classifying land cover have made it difficult to rigorously track global changes in land cover and land use over long time periods. For example, the global area of pastureland reported by

Ramankutty and colleagues (2008) was 18 percent lower than a standard estimate from the United Nations Food and Agriculture Organization because the latter included grazed forestland and semiarid land. However, application of satellite remote sensing, and increasing investments in national level surveys of natural resources, are rapidly improving monitoring capability.

Between 50 and 75 percent of the ice-free land surface on Earth is estimated to be directly altered by humans (Ellis, Goldewijk, Siebert, Lightman, and Ramankutty, 2010; Vitousek, Mooney, Lubchenco, and Melillo, 1997b). Of course, all of Earth's surface is currently subject to alteration from increasing carbon dioxide (CO_2) and climate change linked to anthropogenic greenhouse gas emissions. Key manifestations of land cover change include urbanization, agricultural expansion, and pastureland expansion. The underlying drivers are complex (Ramankutty et al., 2006) but often include demographic factors such as human population growth or migration. Economic integration across the planet associated with globalization (see chapter 7) has markedly influenced land cover and land use in many regions.

Forestland and Shrubland

The spatial extent of global forest area lost is estimated at 15–30 percent of preagricultural forest area (Klein Goldewijk, 2004). About a third of the remaining forests are primary forest, i.e., not directly altered by humans (Morales-Hidalgo, Oswalt, and Somanathan, 2015), much of that in the boreal and tropical zones. The area of primary forest declined 10 percent between 1990 and 2015.

The area of tropical forest (primary plus secondary) continues to decrease, but at a slower rate recently than in the 1990s (R. J. Keenan et al., 2015). There is an ongoing shift in the conversion of natural forests to other uses, with industrial-scale farming, ranching, and commercial forestry (e.g., palm oil) supplanting subsistence farming and grazing. Selective logging and fires degrade large areas of tropical forest even without changing the land cover type.

In the temperate zone, the area of forests has slightly increased in recent years (R. J. Keenan et al., 2015). This "forest transition" refers to a reduction in deforestation that occurs when population size stabilizes and socioeconomic development is successful (Rudel, Defries, Asner, and Laurance, 2009). The timing of that transition in tropical countries with a substantial area of forests remaining may occur earlier in the process now than in the historic past because of rapid urbanization and because forest conservation has become an international issue, with respect to both biodiversity and carbon emissions.

In the absence of artificial disturbance, expansion of tropical rain forests as well as savannah woody vegetation "thickening" is observed in some areas, implying benefits of higher atmospheric CO_2 (Tng et al., 2012). Woodlands are moving into former shrublands in western North America, possible aided by fire suppression and increased water use efficiency from the increasing atmospheric CO_2 concentration (Romme et al., 2007). Some of the more obvious recent changes in global vegetation are in the Arctic region, where climate is changing relatively rapidly. There, increase in shrub cover over large areas is evident in paired photographs from previous decades and today (Provencher-Nolet, Bernier, and Levesque, 2014).

Agricultural and Grazing Land

Roughly one third of global ice-free land is now cropland or pastureland (Ramankutty et al., 2008). Interestingly, the pattern of steady increase in the cropland area throughout the twentieth century has stabilized, while the food supply continues to increase because of agricultural intensification. The global totals in land cover mask continued increases in conversion of forest to agricultural lands in countries such as Brazil (where forests are being cleared for soybean production) and Indonesia (where native forests are being replaced by tracts for palm oil production). The global land cover totals also do not reflect losses in productivity over large areas associated with soil degradation (Gibbs and Salmon, 2015).

Managed grazing land occupies about 30 percent of the terrestrial ice-free surface (Asner, Elmore, Olander, Martin, and Harris, 2004; Steinfeld, Gerber, Wassenaar, Castel, Rosales, and de Haan, 2006). As with cropland, the total global area devoted to grazers is expected to stabilize, while confined systems provide for increases in global livestock production (Thornton, 2010). Much of the ongoing pastureland expansion in Latin

America is associated with deforestation and linked to expansion of cropland into existing pastureland to grow soybeans (Graesser, Aide, Grau,
and Ramankutty, 2015). In many arid regions across the globe, rangeland
is considered overgrazed (Dregne, 1991).

BIODIVERSITY

The survey of Earth system history in chapter 2 indicated multiple episodes of mass extinction. These breaks in the fossil record appear to be
associated with cataclysmic events such as the collision of the Earth with
large meteors, or sustained releases of greenhouse gases from the geosphere into the atmosphere. Five major episodes of extinction have been
identified in the fossil record, and the biosphere is now said to be undergoing a "sixth extinction" (Leakey, 1995). Many more extinction events
are apparent in the fossil record, but the "big five" stand out in terms of
the large proportion (more than 75 percent) of the previously existing
species that went extinct (Barnosky et al., 2011). The current extinction
event differs from the previous episodes in being driven by something
internal to the biosphere. One species (*Homo sapiens*) is responsible for
driving many other species to extinction.

Extinction is the norm in nature: 99 percent of all species that have
existed over geological time have gone extinct. Based on the fossil record,
paleontologists can roughly estimate the rate of extinction over the course
of Earth's history. They find that the rate is episodic rather than continuous, which makes it difficult to compare past extinction rates to the current rate. Estimating the current rate of species extinction is also difficult
because a complete catalogue of species has not yet been assembled. Nevertheless, even crude estimates suggest that if all the species already
driven extinct or formally classified as threatened with extinction (IUCN,
2017) are included, the current trajectory is toward something that could
credibly be termed a mass extinction event (Barnosky et al., 2011). The
recent rate of vertebrate extinctions is 100 times the background rate
(Ceballos et al., 2015).

The sixth extinction has been ongoing for tens of thousands of years.
Some of the early extinctions attributed to humans occurred when
expanding populations of hunter-gatherers encountered animals that had

not evolved in the presence of human predators (e.g., when humans first arrived in Australia and North America). More recently, four new factors have come into play that are driving extinction rates ever higher.

First, technology has increased human effectiveness and efficiency in harvesting wild organisms. A striking example is the harvesting of whales using harpoon guns mounted on motor-driven ships instead of hand-powered harpoons in small boats. Dragging metal nets over the ocean floor by commercial trawlers is more effective than the hook-on-a-line approach to harvesting fish. One person with a chainsaw can cut down a tree that has been growing for a thousand years. Generally, there is no longer much of a contest when humans confront other organisms, except perhaps with regard to our own pathogens.

A second factor in the current extinction crisis is the introduction of exotic species. Relationships among predator and prey, herbivores and vegetation, and especially pathogens and their hosts, are generally fine-tuned by biological evolution. An overly virulent pathogen would kill all its hosts and thus go extinct itself. However, the artificial introduction of an alien species to which local species are not adapted gives the intro-duced species a distinct advantage. In this respect, islands make great case studies. The loss of native bird species in Hawaii is approaching 40 percent, and contributing factors include the intentional introduction of egg predators such as the mongoose (originally introduced to bring down the population of inadvertently introduced snakes), and accidental intro-duction of alien insects that transmit avian viruses (Steadman, 2006). Almost half (42 percent) of the threatened and endangered species in the United States are listed because of human-introduced alien species (Natureserve, 2003).

A third factor in the ongoing wave of extinctions is land cover and land use change with its associated habitat loss (Larsen, 2004). As noted, more than one third of the Earth's terrestrial surface has been converted from native vegetation to agricultural and grazing use, thus largely displacing native species (Ramankutty et al., 2008). On average, 13 million hectares of forestland (mostly tropical forest) have been converted to other uses annually each year in recent years (FRA, 2010). As habitat is lost, animal populations decrease until a minimum can no longer be maintained. Islands like Madagascar, where 90 percent of the forests have been lost and a considerable proportion of the species are endemic, are extreme examples of this factor at work. For nine out of ten species that are considered threatened with extinction globally, habitat loss is the primary driver.

Lastly, on top of direct habitat destruction is the new overlay of habitat loss driven by climate change. Climate is, of course, always changing to some degree. Usually the rates are such that species can track the environmental change by means of dispersal to new areas with climate favorable to their adaptations. A problem now is that humanity has altered the land cover so extensively that species migrations may be blocked, or appropriate geographic areas may have been appropriated by humans for food and fiber production, transportation, or dwelling space. The technosphere has essentially consumed a part of the biosphere, thus limiting prospects for migration.

The current episode of human-induced climate change may lead to very high rates of habitat loss because it is happening relatively fast. We have noted earlier the geologically unprecedented rate of CO_2 increase. Evidence that plant and animal species have already begun to shift their geographic distributions in response to recent climate change (and perhaps increasing CO_2) is beginning to accumulate (e.g., Fei et al., 2017). However, rates of plant species migration—by means of seed dispersal—may not keep up with rates of climate change (see below). In the ocean, organisms may generally be more mobile, but appropriate habitat may be lacking (e.g., for new corals reefs).

Biological communities typically consist of a variety of species that have complimentary roles in maintaining ecosystem metabolism (see chapter 3). Even within a trophic level (e.g., plants), there can be tremendous specialization or niche differentiation with respect to microhabitats (e.g., Inman-Narahari et al., 2014) and regeneration strategy (Grubb, 1977). Conservation of biodiversity is thus rationalized in terms of maintaining ecosystem functionality and resilience (Oliver et al., 2015). The adaptive capacity of ecosystems in the face of climate change will certainly be compromised as species are lost. From a more anthropocentric perspective, loss of biodiversity means less likelihood of discovering useful medicines, foods, chemicals, and genes, as well as potential loss of ecosystem services such as supplying clean water (Seddon et al., 2016).

TERRESTRIAL DISTURBANCE REGIMES AND BIOGEOGRAPHY

Studies of tree mortality (independent of fire and direct human disturbance) have found an upturn in mortality rates in western North America (van Mantgem et al., 2009) and elsewhere (Allen et al., 2010) in recent decades. Proposed mechanisms relate to various manifestations of climate change, notably drought stress. Disturbance agents such as insects and pathogens may also be favored by ongoing climate warming. For example, as seasonal temperatures trend upward, there is a decreased likelihood of achieving minimum killing temperatures in winter, and an increased likelihood of achieving multiple life cycles in summer (Bentz et al., 2010).

The most direct mechanism of tree mortality induced by climate change is an increased incidence of wildfire. Statistical relationships between interannual variation in climate and the incidence of fire support a strong relationship in some regions (Littell, McKenzie, Peterson, and Westerling, 2009). Multiple studies in recent years have documented an increase in the area burned per year in many areas of western North America, generally associated with a warming climate (Westerling, Hidalgo, Cayan, and Swetman, 2006). Globally, the temporal patterns are less clear (Doerr and Santin, 2016), with successful fire suppression a significant factor.

Reviews of the literature with respect to changing geographic distributions of plants and animal species in response to recent climate change find unequivocal evidence that many species have begun moving (Parmesan, 2006). The range shifts are usually northward (in the northern hemisphere) and upslope. The evidence is particularly compelling in the case of the more mobile species, such as birds and butterflies. But even plants, including trees species, are showing evidence of range changes. Notably, the Arctic tree line is advancing to the north in some areas (Hofgaard, Tommervik, Rees, and Hanssen, 2013).

Not all studies of species' distributions are finding the expected latitudinal and elevational shifts. K. Zhu and colleagues (2012) examined data from 43,334 forest plots in the eastern United States that were periodically revisited. Besides noting presence or absence of species, the relative distribution of seedlings and older individuals gives an indication of ongoing range changes. They found no consistent patterns in latitudinal range change among 92 tree species. Some studies are also finding both upslope and downslope range changes, i.e., changes not clearly associated with a "climate footprint" (Rapacciuolo et al., 2014). Complicating factors include the importance of the disturbance regime to regeneration success (Serra-Diaz et al., 2016) and the importance of water balance rather than simply temperature in determining species' distributions (Crimmins, Dobrowski, Greenberg, Abatzoglou, and Mynsberge, 2011).

TERRESTRIAL NET PRIMARY PRODUCTION AND HETEROTROPHIC RESPIRATION

As noted in our earlier discussion of the global carbon cycle (chapter 4), estimates for the proportion of global net primary production (NPP) on land that is diverted to human purposes range from 25–40 percent (Haberl et al., 2007; Krausmann et al., 2013). Given the importance of NPP to human welfare, there is considerable interest in possible trends in regional (Running et al., 2004) and global (Imhoff et al., 2004) NPP, and the proportion of it used by humans. Fortunately, satellite remote sensing is beginning to provide an annual estimate of global NPP based on spatial and temporal patterns in weather and in vegetation greenness (as we will see in chapter 9).

These NPP observations suggest an increasing global terrestrial NPP in the 1989–2000 period (Nemani et al., 2003), a slight decrease in the 2000–2011 period (Zhao and Running, 2010), and an overall small increase over the 1980–2011 period (W. K. Smith et al., 2016). Areas of NPP reduction are often associated with drought, whereas areas of NPP increase are more diffuse. Fertilization and irrigation of croplands have raised NPP over large areas. The long-term trend of globally increasing NPP is probably driven mostly by CO_2 fertilization (Shevliakova et al., 2013; Z. C. Zhu et al., 2016) and climate change (Buermann et al., 2016), but there is ongoing controversy about the relative contributions of those factors (Smith et al., 2016). A readily measurable indicator of the changes in global NPP is the magnitude of the annual drawdown in the atmospheric CO_2 concentration. Especially for observation sites at higher latitudes, such as

Barrow, Alaska, the amplitude of the annual CO_2 oscillation is increasing, suggesting both increased NPP and increased rates of decomposition in these regions of the biosphere (Forkel et al., 2016).

Heterotrophic respiration by decomposing organisms and animals provides a large proportion of the fast carbon cycle return of CO_2 to the atmosphere (see figure 4.1). Since much of the substrate for heterotrophic respiration is provided by NPP, it tends to respond in a similar manner as NPP to climate variation, but with lower sensitivity and a lag (Buermann et al., 2016). Heterotrophic respiration increases with temperature, which introduces sensitivity in the terrestrial carbon sink to climate warming (Anderegg et al., 2015). Fire emissions (wildfire and human caused) are also sensitive to interannual variation in climate in many regions. In relatively warm El Niño years, increased fires in tropical zone forests help drive up the annual increase in atmospheric CO_2 (van der Werf et al., 2008).

The net effect of climate variation on biosphere carbon is that the growth rate in the atmospheric CO_2 concentration increases in relatively warm years (Wang et al., 2013). This creates a concern that the terrestrial carbon sink, which currently helps offset fossil fuel emissions, could decrease as global climate warms.

MARINE PRODUCTIVITY AND BIOGEOGRAPHY

Ocean net primary production is of about the same magnitude as terrestrial NPP, i.e., 50–60 PgC (petagrams of carbon) per year. However, much of it is generated by microorganisms (phytoplankton) rather than multicellular organisms. Ocean NPP provides the base of the marine food web, and possible changes are of great interest, considering the substantial role marine fish play in human protein consumption.

Satellite-borne sensors are used to monitor ocean NPP (see chapter 9), although uncertainties are greater than in the case of monitoring terrestrial NPP. Most reports suggest a decrease in global ocean NPP (D. G. Boyce, Lewis, and Worm, 2010; Roxy et al., 2016; but see Behrenfeld et al., 2016). As on the land, increasing CO_2 concentration in the water could increase ocean NPP (Hays, Richardson, and Robinson, 2005). Increasing sea surface temperatures are detrimental to NPP rates at low latitudes (Nagelkerken and Connell, 2015). As the ocean warms, it tends to stratify, which reduces nutrient upwelling and increases exposure of phytoplankton to harmful ultraviolet radiation. Ocean acidification is also reducing NPP of calcifying phytoplankton (Nagelkerken and Connell, 2015). The production of the algal symbiont in coral is declining because the coral polyps eject the algae when exposed to unusually warm water, which is becoming more prevalent (Baker, Glynn, and Riegl, 2008). Recovery generally occurs as ocean temperatures cool but not in all cases. The estimated rate of Indo-Pacific coral loss is approximately 1 percent per year (Bruno and Selig, 2007).

The most direct influence of the technosphere on the marine biosphere is the harvesting of fish (Pauly, Watson, and Alder, 2005). Unlike on land, where wild-caught protein is no longer a significant proportion of total animal protein consumed by humans, wild-caught fish still constitute a large (about 50 percent), though declining, proportion of total fish consumption (World Bank, 2013a). However, the history of ocean fisheries is a chronicle of overfishing and fisheries degradation. The global catch increased until about 1990, then stabilized. That increase was associated with heavy investment in advanced harvesting technology. The apparent cap was reached because of overexploitation. Approximately 50 percent of global fisheries are fully exploited, whereas 25 percent are overexploited and 25 percent are underexploited (Pauly et al., 2005). Fish production may also be decreasing because of climate change impacts on phytoplankton. For example, there has been a large decline over the past 25 years in Antarctic krill (Atkinson, Siegel, Pakhomov, and Rothery, 2004), which feed on plankton and are consumed by higher trophic level organisms.

As with terrestrial species, marine species are beginning to respond to trends in ocean environmental conditions associated with anthropogenic factors (Parmesan, 2006). Beaugrand and colleagues (2002) reported changes in zooplankton in the North Atlantic over a 40-year period (1960–1999) that were consistent with climate warming, possibly leading to decline of cod, which was already diminished by overharvesting. The species composition of fish assemblages is changing in the North Sea to include more formerly southern species (Beare et al., 2004).

Oceanographers are challenged to evaluate the complete suite of human-induced changes in the ocean environment (Hoegh-Guldberg and Bruno, 2010; Lubchenco and Petes, 2010; Miles, 2009). Add in disruption in coastal areas (Halpern et al., 2008)—including "dead zones" induced by nutrient-rich runoff from agricultural basins (Dybas, 2005)—and it becomes clear that, like the land component of the biosphere, the ocean biosphere is undergoing profound human-induced change.

CHANGES IN POPULATION STRUCTURE AND TROPHIC DYNAMICS

In addition to the direct effects on species extinction rates outlined earlier, there are other more subtle but pervasive human impacts on the biosphere. When harvesting wild organisms, humans typically select the largest individuals they can find. This behavioral pattern has produced shifts in population structure toward smaller sizes and younger ages, a trend that is particularly apparent in fish (Pauly et al., 2005) and trees (Lomalino, Channell, Perault, and Smith, 2001). In the ocean, large body size in vertebrates is associated with an elevated level of extinction threat (Payne, Bush, Heim, Kknope, and McCauley, 2016). The change from primary (native) forest to forest plantations is an ongoing global phenomenon, which tends to truncate the upper end of the age class distribution of associated tree species.

Humans are also disrupting natural processes that regulate populations at the local level. Overhunting and overfishing tend to remove the organisms at the top of the food chain; carnivores on land and large predatory fish in the ocean. A problem then arises because of the loss of the regulatory function of these trophic levels within ecosystems (Ripple and Beschta, 2005). Removal of a regulatory factor (such as a predator species) sets the stage for population explosions that can potentially lead to habitat degradation and a general loss of productivity.

American ecologist Aldo Leopold wrote about the importance of not losing any of the parts as we tinker with natural systems, and the story of deer on the Kaibab Plateau in the southwestern United States is a terrific case study. It helped awaken Leopold to the fallacy of rigid predator control (Ripple and Beschta, 2004). Early in his career as a land manager he did whatever he could to wipe out local populations of coyotes and wolves. However, on the Kaibab Plateau in northern Arizona, he saw that loss of the predator control on the deer population caused a population explosion, which in turn caused severe degradation of the vegetation cover. Eventually the deer population crashed. Leopold came to understand the connections among species in the ecosystem, and it informed his subsequent management approach, which included maintaining populations of the local predators, such as cougars, on the Kaibab Plateau.

A parallel case study in the ocean involves triggerfish, sea urchins, and coral reefs (McClanahan and Shafir, 1990). In Kenyan reef lagoons, triggerfish normally control the sea urchin population, and sea urchins partially regulate coral reefs by their influence on bioerosion. When the triggerfish are overexploited, the sea urchin population rises, and coral reefs are degraded.

Another concern with respect to relationships among trophic levels is that climate change–induced alteration of species ranges and phenology is resulting in asynchrony between the period of high primary production and the presence of consumers (Visser and Both, 2005). On land, there is concern that the insect pollinators could respond differently to climate change than their usual floral hosts. In the ocean, secondary production is often dependent on synchronous plankton production (Edwards and Richardson, 2004).

With respect to human harvesting of biosphere productivity, the usual step on land beyond replacing the top predators with ourselves is to replace the herbivores with a domesticated variant. As with fisheries management, the history here is rather bleak. Classic books such as *Topsoil and Civilization* (Dale and Carter, 1955) document the long-term effect of overgrazing in the Middle East on the land and biota. Likewise, in the America West, herds of cattle and sheep swept over the land in the late 1800s and had strong and lasting impacts on the distribution and productivity of many grass species, notably spring season bunchgrasses (Abruzzi, 1995).

A last pervasive but subtle influence of the technosphere on the biosphere is the inadvertent dispersion of environmental contaminants (UNEP, 2012). Broadly characterized as endocrine disrupters, the chemicals are used in industrial, agricultural, and pharmaceutical applications (Schug et al., 2016). Broad public awareness of a problem began with Rachel Carson's *Silent Spring* in 1962, and the scientific understanding of the issues has progressed significantly since then (Guillette and Iguchi, 2012). However, identifying the relevant metabolic mechanisms (often related to functioning of hormones) and characterizing the degree of an organism's exposure to the chemicals in its native environment, make environmental toxicology an especially complex area of research and policy development.

IMPACTS ON GENE POOLS

In many ways, human domination of ecosystems is reducing global genetic diversity. We have noted the current high rate of species extinction, and consequent loss in species diversity. Also contributing to the overall loss in genetic diversity is reduction in the geographic range of many species associated with overharvesting, land use change, and climate change. Widely distributed plant and animal species often evolve locally adapted populations based on genetic differentiation. Hence a loss in the geographic range means a loss in genetic variation within a species. Genetic diversity of agricultural species is also being lost as industrial agriculture focuses on a limited number of domesticated animal and crop species (Khoury et al., 2014), as well as a narrow range of genotypes within each domesticated species. This simplification of the biosphere is contrary to the usual trend in biological evolution that tends to promote functional specialization and genetic adaptation to local conditions.

The increasing prevalence of disturbed environments means that we are creating a biosphere dominated by more weed-like species. There will be fewer highly specialized life forms with life cycles finely tuned to their environment, and a greater prevalence of generalist forms that thrive on disturbance. There will be less redundancy within the guilds of species that are components of ecosystems. In this respect, we are moving away from what may be needed to sustain ecosystem services in a rapidly changing world (Loreau et al., 2001).

Interestingly, we are creating more genetic diversity in one sense. Through traditional breeding programs, and more recently with genetic engineering, we are adding new elements to the global gene pool. Domesticated plants are genetically altered to withstand commercial herbicides and resist specific pests and pathogens; the biochemistry of microbes is genetically altered so they produce useful compounds like insulin; and fish genes are manipulated to increased productivity under artificial conditions. Nature is of course responding; we have witnessed the appearance of herbicide-resistant weeds and antibiotic-resistant bacteria. The technosphere has entered a genetic arms race with pests and pathogens.

Most broadly, we are slowly erasing the original gene pool extant at the beginning our ascendancy about 50,000 years ago, and replacing it with new genotypes that are adapted to our wants and needs. Evolution is increasingly driven by artificial selection (human influenced) rather than classical Darwinian natural selection.

BIOSPHERE SURVIVAL

The survival of the biosphere in the face of the current technosphere-driven perturbation is not in doubt. Much of biosphere metabolism is microbial, and Earth history suggests that the microbial component of the biosphere can withstand even a large reduction in the number of higher order species (e.g., 96 percent of all marine species and 70 percent of all terrestrial vertebrate species, as occurred in the end-Permian mass extinction). Stressed ecosystems (e.g., a polluted lake) often degrade into a state of lower biodiversity and energy throughput (Rapport and Whitford, 1999). Hence, a biosphere stressed by technosphere interventions will likely persist, but for human purposes it will be less hospitable than Earth's biosphere as our ancestors knew it, or even as we know it today.

ESSENTIAL POINTS

We are rapidly converting natural capital in the form of biodiversity, biomass, and functional ecosystems into technosphere capital in the form of cities, transportation infrastructure, and telecommunications devices. The impoverishment of communities and ecosystems goes beyond declining biodiversity to include changes in population structure and genetic variation. Global terrestrial net primary production is apparently increasing, mostly in response to a rising CO_2 level, whereas marine net primary production is apparently decreasing in association with ocean warming, stratification, and acidification. Disturbances are a usual feature of ecosystems, but disturbance regimes are now changing because of both direct human impacts associated with resource utilization and indirect human impacts by way of climate change.

IMPLICATIONS

1. The irreversible loss of biodiversity owing to human actions will lead to a planet that is less hospitable for our species.

2. The massive losses of biodiversity predicted for the future will diminish the resilience of the associated ecosystems.

3. If biosphere carbon sequestration begins to decrease instead of continuously increase, the prospects weaken for stabilizing and reducing the atmospheric CO_2 concentration.

FURTHER READING

Morales-Hidalgo, D., Oswalt, S. N., and Somanathan, E. (2015). Status and trends in global primary forest, protected areas, and areas designated for conservation of biodiversity from the Global Forest Resources Assessment 2015. *Forest Ecology and Management, 352*, 68–77.

Williams, M., Zalasiewicz, J. A., Haff, P. K., Schwagerl, C., Barnosky, A. D., and Ellis, E. C. (2015). The Anthropocene biosphere. *Anthropocene Review, 2*, 196–219.

6

SCENARIOS OF GLOBAL ENVIRONMENTAL CHANGE

Society may be lulled into a false sense of security by smooth projections of global change . . . present knowledge suggests that a variety of tipping elements could reach their critical point within this century under anthropogenic climate change.

—T. M. Lenton et al. (2008)

anaging a natural resource (in this case the Earth system) is optimally based on monitoring—to understand status and trends—as well as an ability to forecast the response of the resource to specified conditions. Such conditional predictions (scenario development) at the global scale serves two distinct functions (van Vuuren et al., 2011). With respect to the climate change issue, scenarios provide decision support for international negotiations about agreements to reduce greenhouse gas emissions. At the national level, mitigation investments (such as subsidizing renewable energy development) compete with other societal spending needs and must be justified in terms of portraying the consequences of inaction and the benefits of actions. Forecasting can also be used to develop climate change adaptation plans; e.g., expectations about future sea level rise and flooding risk can inform contemporary policies on land use. These types of societal decisions must inevitably be made in the context of uncertainty about the future, and scenario development provides an approach to managing that uncertainty (Parson, 2008).

A science fiction writer can, of course, readily create a scenario for the near-term or long-term future of the Earth system. However, Earth system scientists go a step beyond that by constructing computer-based simulation models of Earth system processes. The beauty of this approach lies in building into the models what we know about the physical and biological mechanisms that drive the Earth system, e.g., the physics of how greenhouse gases affect the climate. Besides producing well-informed scenarios of the future, simulations provide the opportunity to test the implications of alternative assumptions about the future, e.g., the rate of growth in the atmospheric carbon dioxide (CO_2) concentration.

A process model that simulates a component, or all, of the Earth system is basically an assemblage of algorithms (sets of equations) that quantitatively specify changes in states (e.g., air temperature) and process rates (e.g., rainfall) through time. A process model is a simplification of reality, with necessary compromises regarding the level of detail that is included in the algorithms. In building a simulation model, complex research issues arise concerning what processes to include and at what level of detail, what temporal resolution is optimal, and what spatial resolution is appropriate if the model is spatially explicit.

Development of Earth system models benefits significantly from observations at a range of spatial and temporal scales. Model calibration is the effort to determine the values for model parameters or coefficients. For example, leaf-level measurements of transpiration (water flux) with a cuvette that measures gas exchange help specify the controls on canopy transpiration in forests. Model evaluation is concerned with testing the accuracy of model outputs. Global climate models can be tested with observations in the historic record. The sensitivity of the climate system to greenhouse gas concentrations is also inferred from earlier eras in the geological record (J. Hansen et al., 2013; Pagani, Liu, LaRiviere, and Ravelo, 2010), including the ice core–based estimates of the oscillation in atmospheric CO_2 concentration and global mean temperature over the past 800,000 years.

Beyond their role in climate change mitigation and adaptation, Earth system models are used to develop and test hypotheses about how the Earth system works (Prinn, 2013). Experimental model runs can be made to isolate the role of specific processes in regulating climate (e.g., the role of anthropogenic aerosols in climate cooling).

In this chapter, we examine the tools used in Earth system scenario development, and review key indicators commonly featured to characterize the future state on Earth. Five synoptic scenarios covering a range of Earth system futures are described.

GLOBAL CLIMATE MODELS

The physics of what is commonly called the greenhouse effect, associated with atmospheric trace gases like CO_2, was understood by the late 1800s (see chapter 2). By the late 1900s, scientists had built three-dimensional (3-D) models of the atmosphere to simulate the consequences of increased greenhouse gas concentrations on the climate system as a whole (figure 6.1). These general circulation models (GCMs) simulated horizontal and vertical transfers of mass and energy for a 3-D grid of cells surrounding the Earth. The energy and water balance at the surface of the Earth was an essential feature of the models.

A common GCM experiment was to calibrate the model to replicate the contemporary global climate given the contemporary CO_2 concentration. Then the concentration of CO_2 in the atmosphere was doubled (e.g., from 280 to 560 ppm [parts per million]). After allowing the model to reach equilibrium, the new climate was characterized, and differences from the 280 ppm climate assessed. This simple configuration allowed for fast feedbacks in the climate system, e.g., the fact that higher air temperatures led to higher water vapor concentrations over the oceans, which means more greenhouse gas–induced warming. However, it did not account for several slow feedback mechanisms, e.g., ocean warming, and melting of the Greenland and Antarctic ice sheets, which combined could easily double the overall climate sensitivity (i.e., the long-term change expected from a doubling of atmospheric CO_2).

The initial GCMs had what was known as a swamp or slab ocean. It exchanged heat and moisture with the atmosphere, but did not account for heat transfer horizontally (e.g., currents) or heat transfer vertically to the interior ocean. Building a spatially explicit dynamic ocean simulation model, with multiple layers, was a major challenge because of factors that included the limited observations needed for model initialization and calibration, the slow response times of ocean processes, and the

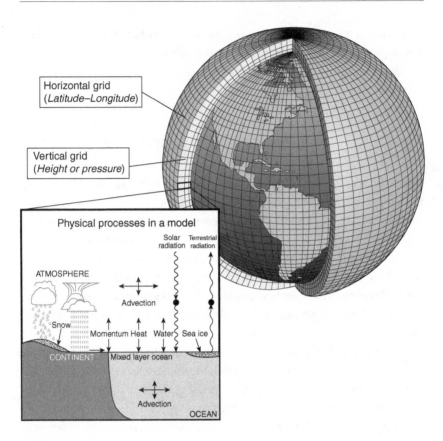

FIGURE 6.1

Three-dimensional grid of the earth exemplifying coupled atmosphere–ocean general circulation models. Climate models use differential equations from physical laws (e.g., fluid motion) and chemical relationships, calculated on supercomputers. The box shows some of the major components in climate models. Grid dimensions (latitude/longitude, vertical grid height/ocean depth) vary among models. Courtesy of the National Oceanic and Atmospheric Administration. Figure and caption used by permission from Cronin (2009).

requirement for higher spatial resolution than is needed in models of the atmosphere. However, it was necessary to build coupled ocean-atmosphere models to begin capturing the temporal dynamics of projected climate change, which are important in a policy context.

A notable uncertainty in GCMs is the sign and magnitude of the cloud feedback to climate warming. Intuitively, warmer temperatures would lead to more atmospheric moisture, which could lead to more clouds, hence more reflection of sunlight back to space and a negative feedback to warming. However, there are many kinds of clouds, and the energy balance of the atmosphere is quite complex. GCMs typically operate at spatial resolutions (horizontally and vertically) that do not resolve the physics of actual cloud formation and dissipation; thus, various approaches have been taken to simulate cloud dynamics. The variation among GCMs in their treatment of clouds is a large contributor to their variation in climate sensitivity to greenhouse gas increases (Cess et al., 1996; Zhao et al., 2016). Interestingly, recent observations (Trenberth, Zhang, Fasullo, and Taguchi, 2015), and comparison of GCM simulations to observations (Clement, Burgman, and Norris, 2009), suggest an overall positive cloud feedback (i.e., amplifying) to climate warming.

Besides checking climate model outputs against observations in the historical record, GCMs have been run with conditions of CO_2 concentration, land mass configuration, and solar radiation in the paleorecord, and tested against inferred climate at the time. As noted, climate sensitivity in the paleorecord is a check on the climate sensitivity in the models. In comparing the two, researchers look at the equilibrium temperature change per unit of forcing in terms of watts per square meter. Paleoclimate studies support a range of climate sensitivities—some lower than (Schmittner et al., 2011) and others greater than (J. Hansen et al., 2013) the midrange seen in CMIP5 (Taylor, Stouffer, and Meehl, 2012), the most recent climate model intercomparison exercise. (CMIP5 is an acronym for the fifth phase of the Coupled Model Intercomparison Project, which provides a forum for global climate model evaluation and improvement.) The paleorecord also indicates a trend of increasing climate sensitivity as Earth warms (Shaffer, Huber, Rondanelli, and Pedersen, 2016).

GLOBAL INTEGRATED ASSESSMENT MODELS

In early climate models, the greenhouse gas concentrations were prescribed and the climate was allowed to equilibrate. However, as noted in chapters 2 and 3, there are multiple slow feedback loops within the climate system. It is thus necessary to run a climate model in the "transient" mode, e.g., from year 2000 to 2100, to fully capture the impacts of increasing greenhouse gas concentrations. Consequently, a new requirement for running a $2 \times CO_2$ GCM experiment (e.g., 560 ppm versus 280 ppm) became a projection for the changing greenhouse gas concentrations over time for input to the climate model. This approach called for modeling fossil fuel emissions and land use change emissions as affected by factors such as population growth and economic development, i.e., modeling the global economy and human use of natural resources.

The history of prognostic (predictive) modeling of the global economy and associated natural resources use is short. Early efforts were made by the Club of Rome group in the 1970s. The Club of Rome was a nongovernmental organization (a think tank) founded by European businessmen interested in the global future. Their work generated a surge of attention to the global-scale perspective (Meadows, Meadows, Randers, and Behren, 1972). The modeling used what would now be considered primitive computers and algorithms to simulate the interactions of humanity with the global environment.

The Club of Rome modeling approach was not spatially explicit, and it operated at an annual time step. Assumptions were made about demographic parameters, such as the rates of population increase, as well as per capita resource use, rates of economic growth, levels of available resources, per capita pollution, and limits to the Earth system's capacity to absorb pollution. Global climate change was not considered.

The original Club of Rome scenarios painted a rather bleak trajectory for humanity—generally a trend of exponential population and economic growth for a few decades, and then a population crash associated with exhaustion of natural resources. The book that summarized the scenarios (*The Limits to Growth*) sold well (30 million copies), but the modeling was heavily criticized, particularly by economists. Famously, economist Julian Simon bet ecologist Paul Ehrlich that the inflation-adjusted price of a set of five strategic metals would not rise over the course of the 1980s. It turned out that Simon won, thus lending support to the believers in the technological fix (i.e., the idea that advances in technology will ultimately solve environmental problems associated with economic growth).

Beginning about 1990, a new generation of global-scale integrated assessment models (IAMs) arose. Climate change was now very much in the picture. These IAMs usually ran over a grid covering the planetary surface and included simplified simulation of the climate and the biogeochemical cycles, as well as the economy. Demographic and economic submodels were run at the global or regional scale (IMAGE, 2017). An important interaction that was captured was effects of population growth on deforestation, and hence related effects on CO_2 emissions associated with land use change (see chapter 4). Assumptions were made about climate sensitivity to greenhouse gas concentrations, and economic damages associated with changes in global mean temperature (i.e., damage functions). A key output of this new generation of global models was an estimate for the trajectory of the atmospheric CO_2 concentration as influenced by fossil fuel emissions and deforestation. In some scenarios, carbon sinks were assumed based on afforestation; in others, negative emissions were driven by bioenergy facilities coupled with carbon capture and burial. Most generally, these models showed that carbon emissions from projected fossil fuel combustion and deforestation over the twenty-first century would drive up the concentration of CO_2 to levels that would significantly alter the global climate and the global economy.

Important drivers of CO_2 emissions in IAMs include the following.

- *Global population and per capita resource demand.* Generally, as these variables go up, the projected CO_2 emissions rise. Growth in population and in labor productivity are usually prescribed based on internally consistent storylines about socioeconomic development (O'Neill, Jiang, and Gerland, 2015; Samir and Lutz, 2014).
- *GDP growth.* The global summed gross domestic product (GDP, in constant year 2000 dollars) is a measure of total economic activity. GDP emerges from assumptions about populations and economic growth. High GDP leads to higher greenhouse gas emissions.
- *Partitioning of energy sources.* This partitioning generally refers to the degree to which fossil fuel energy generation is replaced by non–fossil fuel sources (nuclear, wind, solar, hydroelectric, biomass, geothermal). Modeled shifts in energy technology are driven by market competition and hence influenced by factors such as a carbon tax on fossil fuels. Scenarios with a rapid global shift toward renewable energy technologies result in lower CO_2 concentrations by 2100.

- *Land use and land cover change.* Rates of deforestation and afforestation affect the net land source or sink for carbon. These rates are influenced by assumptions about population growth and urbanization (Alcamo et al., 2006).

Advanced IAMs now treat all the major greenhouse gases, including methane and nitrous oxide emissions from agriculture.

Besides projecting greenhouse gas emissions, IAMs are used by economists to evaluate alternative policy options for mitigating climate change. As we have observed, greenhouse gas emissions are for the most part consider an externality—everyone burns fossil fuels without regard to their possible negative impacts on the environment. However, we have been awakened to the fact that greenhouse gas–induced climate change will have significant costs. Economist are thus charged with determining the "social cost of carbon" (i.e., how much an emitter should pay to ameliorate the damages associated with climate change). Collecting this cost at the time of carbon emissions—as with a carbon tax—would potentially begin altering the partitioning of energy sources. As one might expect when economic consequences are large, this endeavor has been controversial (Ackerman, DeCanio, Howarth, and Sheeran, 2009).

EARTH SYSTEM MODELS

Rapid technological improvements for storing and processing data have made it possible in recent decades to run increasingly complicated global climate models at increasingly finer spatial resolution, and over longer simulation periods (McGuffie and Henderson-Sellers, 2005). However, that increasing complexity does not guarantee reduction in the uncertainty of the projections. For instance, the range for projected increases in mean global temperature associated with a doubling of CO_2 has been increasing rather than decreasing (Freeman, Wagner, and Zeckhouser, 2015). Nonetheless, an increase in model complexity usually correlates with an improved understanding of the system being modeled, and perhaps less likelihood of missing a critical process.

The most comprehensive global climate models are classified as Earth system models (ESMs). Although they vary in the processes covered or emphasized, These models increasingly include the following components (Kump, Kasting, and Crane, 2010).

1. The horizontal and vertical circulation of the atmosphere—including formation and dissipation of clouds, and the configuration of large-scale features like the Hadley cell, the jet streams, and the Walker circulation.

2. The circulation and vertical mixing of the ocean, including the thermohaline circulation (THC).

3. The energy balance of the atmosphere, land, and oceans. Relevant atmospheric phenomena include radiation transfer associated with greenhouse gases, aerosols, clouds, and energy transformations associated with water condensation/evaporation.

4. The cryosphere, i.e., the formation and melting of snow, sea ice, glaciers, and ice caps.

5. The carbon cycle on unmanaged lands (and in some cases the nitrogen cycle as well), i.e., primarily the fast carbon cycle (photosynthesis, biomass accumulation, mortality, and decomposition). Unmanaged lands can be a source or sink of CO_2 depending on factors such as CO_2 fertilization and the incidence of fire (Cox, Betts, Jones, Spall, and Totterdell, 2000).

6. The carbon cycle and energy balance consequences of land cover and use change. The geographic distribution of different land cover types influences the land surface energy balance and fluxes of greenhouse gases. Hence in recent ESM intercomparisons such as CMIP5, a representation of future land cover (e.g., the area of agriculture) is prescribed (Hurtt et al., 2011).

Besides the many independent teams developing ESMs, the U.S. National Center for Atmospheric Research (NCAR) maintains and coordinates development of a Community Earth System Model (CESM, 2017) and a Community Land Model (Lawrence and Fisher, 2013). The computer code for these models uses modular programming so individual researchers can work to improve specific algorithms. Periodically, a meeting is called and a consensus is reached about upgrading specific components of the model based on recent research.

KEY INDICATORS IN EARTH SYSTEM SCENARIOS

An Earth system scenario is an internally consistent projection of how the planet will change over a given period. IAMs are used to tie together assumptions about factors such as population increase, economic growth, technological advances, and policy commitments to reduce carbon emissions. IAM projections for concentrations of greenhouse gases are then input to ESMs that simulate climate change (although models that simulate the carbon cycle can also accept emissions themselves as input). Some of the key outputs of ESMs with respect to understanding Earth's future are as follows.

- *Air temperature.* The mean global air temperature is a common indicator of climate change. Its current value is at about 14.8°C, having risen approximately 0.8°C over the past century. An interesting reference point to climate model projections of global mean temperature change is the change in global mean temperature associated with the Pleistocene ice age cycles (see chapter 2). Based on the ice core record of stable isotopes, it is estimated that global mean temperature was 4°C–7°C cooler during the glacial maximums compared to the glacial minimums (IPCC, 2014b). The high end of the recent climate change simulations suggests an increase of similar magnitude by the year 2100, with more warming to follow in subsequent centuries.

- *Atmospheric circulation.* As the climate warms, the jet streams (storm tracks) begin to meander more, and increasingly to get stuck in certain configurations. The net effect is a greater likelihood of extreme weather. The greater amplitude of the meanders means Arctic air penetrates farther south on occasion, and likewise warm temperate zone air penetrates farther north. Getting stuck means there are longer periods of the same kind of weather in a given place (e.g., longer droughts). The multiyear drought in California in recent years is associated with a configuration of the jet stream such that precipitation-laden storms were displaced to the north (S. Y. Wang, Hipps, Gillies, and Yoon, 2014). The jet streams may also be displaced toward the poles, which reduces precipitation in some mid-latitude areas (Chang, Guo, and Xia, 2012).

- *Ocean circulation.* The vigor of the THC (introduced in chapter 2) is indicated by the volume of water sinking in the North Atlantic. The North Atlantic component of the THC is termed the *Atlantic meridional overturning circulation* (AMOC). There has been an indication of declining AMOC volume in the twentieth century (Rahmstorf et al., 2015). In coupled ocean-atmosphere climate models, AMOC slows over the course of the twenty-first century (Weaver et al., 2012), in some cases essentially stopping in high global warming scenarios (Prinn, 2013). The consequences include more global climate warming because less heat is mixed into the ocean interior, and less CO_2 uptake by the ocean because the CO_2 concentration is high in cold North Atlantic water when it sinks. The northern hemisphere could cool significantly if AMOC slows (Vellinga and Wood, 2008). More generally, as the THC slows, the ocean becomes more stratified.

Another large-scale ocean circulation feature with significant climate implications is the Walker circulation, associated with the El Niño–southern oscillation (ENSO). Here, surface water flows east to west across the surface of the Pacific Ocean in the equatorial region, and returns west to east at greater depths (NOAA, 2017b). This circulation periodically slows (El Niño years) or speeds up (La Niña years), with the consequence of lesser or greater upwelling of cool deep water off the coast of South America. Global climate is broadly disrupted on these occasions (NOAA, 2017b). ESMs have limited capacity to simulate ENSO processes and vary widely in outcome with respect to greenhouse warming impacts on Walker circulation. Although ENSO extremes are expected to increase (W. J. Cai et al., 2015), there is limited consensus among climate modelers about long-term changes in this important feature of global climate.

- *Ocean acidity.* Current ocean pH (a measure of hydrogen ion concentration) has measurably changed (i.e., acidified) in recent decades. The hydrogen ion concentration near the surface has risen about 30 percent since 1900 and will continue rising as the CO_2 concentration increases (Caldiera and Wickett, 2005), with considerable biological effects (Hall-Spencer et al., 2008).

- *Sea level rise.* Global sea level rise is often quantified as relative to the preindustrial level. It has risen 100–200 millimeters since that time.

Under the most extreme global warming scenarios, sea level could rise as much as 1 meter in the next 100 years, and more than 50 meters in the next 10,000 years (P. U. Clark et al., 2016). Globally averaged sea level rise is driven primarily by melting of glaciers and ice caps, along with thermal expansion of the ocean.

- *Cryosphere dynamics.* Arctic polar sea ice is already in retreat, which as noted earlier is part of a positive feedback loop to climate warming. The decline of the Greenland and Antarctic ice caps contributes to sea level rise. The gradual decline in mountain snow and glaciers reduces the supply of stream flow to lower elevations in the summer.

- *Carbon cycle behavior.* The global carbon cycle has received tremendous attention from the ESM community because of its linkage to global climate.

Typically, ESM projections that include the terrestrial carbon cycle show a carbon sink in the early twenty-first century associated with CO_2 enhancement of photosynthesis, but transition to the status of carbon source later in the twenty-first century. The mechanisms of that late-century positive feedback to climate warming are extensive forest fires and rapid rates of decomposition (a source of CO_2) in the soil. The extra carbon emissions toward the end of the century contribute to raising the CO_2 concentration, thus putting additional pressure on efforts to reduce fossil CO_2 emissions. The ocean carbon sink is influenced by ocean temperature (the solubility pump) and the THC (see chapter 4). Given the high concentrations for atmospheric CO_2 in the future, CO_2 will likely continue diffusing into the ocean surface.

Additional indicator variables can be evaluated with "off-line" model runs in which an ESM output, usually climate or CO_2, is used to drive a simulation model (an "impacts" model) for another variable that depends on climate (Arnell and Gosling, 2016). In the off-line model runs, there is no feedback to the ESM.

- *Vegetation distribution.* Over geological history, global biogeography has responded to changing climate (Yu, Wang, Parr, and Ahmed, 2014). As the climate warms in the twenty-first century, particular ecosystem types or biomes will move upslope and north (in the northern hemisphere) to remain within a climatic envelope of favorable conditions. Biogeographers have studied the relationships of vegetation type to climate and built predictive models for how regional and global vegetation will respond to climate change. In the simplest modeling approach, a statistical relationship is established between the current climate and vegetation distribution (e.g., boreal forests occur where mean annual temperature is within a certain range). With those statistical relationships, and new gridded climate projection produced by an ESM, a new global vegetation distribution can be projected. More sophisticated dynamic global vegetation models (DGVMs) also account for the disturbances (e.g., fire) needed for one vegetation type to displace another, and for factors such as seed dispersal rates, which might constrain vegetation redistribution (Bachelet and Turner, 2015).

• *Biodiversity*. Global biodiversity is already under siege in the early twenty-first century and will decline further as human impacts on the Earth system become increasingly manifest (see chapter 5). However, the prospects for modeling changes in biodiversity are limited. One effort (Jantz et al., 2015) took the global gridded land use projections used in the Intergovernmental Panel on Climate Change scenarios (described next) and juxtaposed them with information on current hotspots of biodiversity (often areas of many endemic species). Results indicated losses of 26–50 percent of natural vegetation cover in the hot spot areas. Those area losses, along with local studies of species per unit area, suggest losses of 0.2–16 percent of local species by 2100. As noted in chapter 5, the reality will likely be much worse because of the influence of climate change on habitats.

• *Agriculture*. Crop production is highly dependent on climate, and impacts of climate change on global agricultural production are of great concern. Crop models may be empirical or process based and are calibrated and validated with observations of cropland production. Some of the most important considerations are temperature effects on photosynthesis and respiration, benefits of increasing CO_2 concentration on photosynthesis and water use efficiency, changes in phenology, and possible needs for more irrigation (IPCC, 2014b).

• *Hydrologic impacts*. Hydrologic modeling juxtaposed with population projections give an indication of the number of people exposed to water scarcity, river flooding, and coastal flooding.

- *Changes in energy demand for heating and cooling.* These projections are based on heating and cooling degree days, population sizes, and assumptions about heating and cooling technology.

THE IPCC SCENARIOS TO 2100

The Intergovernmental Panel on Climate Change (IPCC) is charged by the United Nations with periodically summarizing scientific information on climate change (see chapter 8). Projections of climate change impacts based on GCMs, IAMs, ESMs, and impacts models are a critical component of the IPCC deliberations. However, there are 20 or more climate modeling teams hosted by a variety of academic and governmental institutions that provide climate change projections, and a wide variety of impacts models.

To enable direct comparisons of their outputs, IPCC has specified (for the fifth assessment, AR5), several representative concentration pathways (RCPs), designated by numbers ranging from 2.6 to 8.5. These numbers refer to the additional watts per meter squared (equivalent to a small proportion of solar input) introduced to the atmosphere by the end of the twenty-first century by a prescribed increase in greenhouse gas concentrations (Meinshausen et al., 2011).

The greenhouse gas concentration pathways are based on IAMs, described earlier. The essential output from the IAMs in this context is the concentration pathway over the twenty-first century for the important greenhouse gases as influenced by anthropogenic factors, including policy decisions about emissions mitigation. These concentrations become inputs to the climate models. The outputs of climate models are typically summarized in the form of changes in mean global temperature by 2100, with a quantitative indication of uncertainty (box 6.1). Validation exercises with historical climate have suggested that mean output values across a set or "ensemble" of climate models is usually closer to observed values when compared with results from individual models making up the ensemble (Peng, Kumar, van den Dool, and Barnston, 2002). Thus, mean values across an ensemble of models and the variation across models are commonly reported.

BOX 6.1
Climate Model Uncertainty

Nobody knows the future. Thus, any projection could be right or wrong. In the case of complex global climate simulation models, the projections scientists churn out could be right in some ways but wrong in others. At the same time, policy makers crave certainty. A restricted planning horizon, range of potential outcomes, and modest degree of uncertainty about those outcomes makes it more manageable to devise appropriate policy responses to opportunities and potential threats.

Climate modelers work to minimize uncertainty in a specific model by running it for current climate conditions and using observations to calibrate and test the model. Uncertainty in a given model is reported in a formal statistical manner. However, to give an indication of uncertainty in model projections, a set of climate models are run with the same or similar inputs (i.e., boundary conditions and forcings). Results from such an ensemble of models are reported as a mean across models with an indicator of variability among the models. Given an increase in atmospheric CO_2 concentration to 560 ppm, we might find a range of climate sensitivities (i.e., the expected change in global mean temperature) on the order of 2°C–4°C, with a mean of 3°C and a standard deviation across models of 0.5°C.

We must recognize that the variability here is not a measure of uncertainty about the real answer (which cannot be known). Rather, it is an indicator of how much agreement there is among the models about the answer. They might all be biased in a similar way, in which case the uncertainty would appear to be low. The paleorecord of the climate, as well as the historical record, offer opportunities for testing based on measurements, but here, of course, the uncertainty of the measurements is also an issue. Increasingly, an analysis framework that accounts for uncertainties in both the observations and the models is used in model development and assessment.

The most recent Intergovernmental Panel on Climate Change report adopted an approach to characterizing uncertainty with two components (IPCC, 2015, Box TS.1). Confidence in the validity of a finding, i.e., based on "type, amount, quantity and consistency of evidence," is expressed qualitatively. High agreement (confidence) would be assigned to a finding based on robust evidence. When quantitative evidence (i.e., with a statistical basis for uncertainty), is available, a probabilistic approach is taken (e.g., a finding that is deemed "very likely" has a 90–100 percent probability of occurring).

For the most recent (fifth) IPCC assessment, four RCP scenarios were developed (IPCC, 2014a). Each was formulated at a different national laboratory, using a different IAM. Together, they represent a wide array of assumptions about future fossil fuel emissions, population increase, economic development, and patterns of land use for the rest of the twenty-first century (figure 6.2, table 6.1). The IPCC makes no statement about the relative likelihood of the different scenarios, but the RCPs are arrayed in a best case to worst case fashion with respect to induced climate change by 2100.

Business-as-Usual (BAU) Scenario (RCP 8.5)

The BAU scenario assumes growth in the global population and economy are maintained throughout the twenty-first century, and that limited effort is made to constrain fossil fuel emissions. Coal combustion remains a dominant source of energy. Per capita income growth is small, and there is little convergence between rich and poor countries. CO_2 emissions grow throughout the century to reach 27 gigatons of carbon (GtC; 1 Pg = 1 Gt = 10^{15} grams) per year by 2100. CO_2 concentration rises to 950 ppm. As we saw in chapter 2, CO_2 concentrations of around 1,000 ppm have not been seen on Earth for tens of millions of years.

In the ESM outputs for the BAU scenario, global mean temperature for 2100 rises by about 3.7°C (averaged across climate models). Sea level rises about 0.6 meters. In the ocean, pH decreases (0.30 pH units) and AMOC

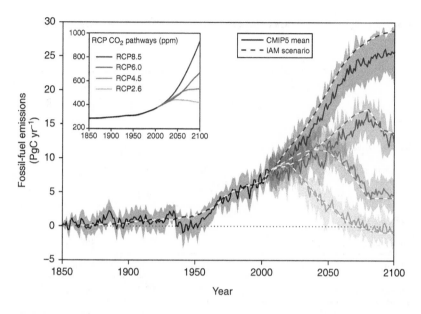

FIGURE 6.2

Historical and projected fossil fuel emissions (main figure) and concentrations (insert box) for Intergovernmental Panel on Climate Change (IPCC) representative concentration pathway (RCP) scenarios (Stocker et al., 2013). CMIP5 = Fifth Coupled Model Intercomparison Project; IAM = Integrated Assessment Model.

Used by permission from IPCC (2013). Stocker, T. F., D. Qin, G.-K. Plattner, L. V. Alexander, S. K. Allen, N. L. Bindoff, F.-M. Bréon, J. A. Church, U. Cubasch, S. Emori, P. Forster, P. Friedlingstein, N. Gillett, J. M. Gregory, D. L. Hartmann, E. Jansen, B. Kirtman, R. Knutti, K. Krishna Kumar, P. Lemke, J. Marotzke, V. Masson-Delmotte, G. A. Meehl, I. I. Mokhov, S. Piao, V. Ramaswamy, D. Randall, M. Rhein, M. Rojas, C. Sabine, D. Shindell, L. D. Talley, D. G. Vaughan, and S.-P. Xie, 2013 Technical Summary. In *Climate Change 2013: The Physical Science Basis. Contribution of Working Group I to the Fifth Assessment Report of the Intergovernmental Panel on Climate Change* [Stocker, T. F., D. Qin, G. K. Plattner, M. Tignor, S. K. Allen, J. Boschung, A. Nauels, Y. Xia, V. Bex, and P. M. Midgley (eds.)] (Cambridge University Press, 2013), 33–115, doi:10.1017/CBO9781107415324.005. https://www.ipcc.ch/pdf/assessment-report/ar5/wg1/WG1AR5_TS_FINAL.pdf.

TABLE 6.1 **Key Assumptions and Outputs from IPCC Scenarios for 2100**

	REFERENCE YEAR 2000	IPCC SCENARIO		
		RCP2.6[a]	RCP 4.5[b]	RCP 8.5[c]
IAM variables				
Population (billion people)	6.0	8.5	8.7	12
CO_2 (ppm)	375	425	525	950
GDP (2,000$)		330	325	250
Oil consumption (Ej/yr)	160	50	180	160
Primary energy consumption (Ej/yr)	400	800	1,000	1,700
Energy intensity (Gj/$)	15.5	2.5	3.0	7.6
Carbon emissions (GtC/yr)	7.5	−3.2	3.6	27.0
Cumulative 21st-century carbon emissions (GtC)		500	820	2200
ESM variables				
Global mean temperature increase (˚C) (range across models in parentheses[d])	0.8	1.0 (0.3–1.6)	1.8 (1.1–2.6)	3.7 (2.6–4.8)
Sea level rise (m)[d]	0.15	0.40	0.47	0.63
AMOC decrease[e] (range in percentage)			5–40	15–60
Ocean pH decrease[e]		0.06	0.14	0.30

Note: AMOC = Atlantic meridional overturning circulation; ESM = Earth system model; Ej = exajoule; GDP = gross domestic product; Gj = gigajoule; GtC = gigaton of carbon; IAM = integrated assessment model; IPCC = Intergovernmental Panel on Climate Change; RCP = representative concentration pathway.

[a] Calvin et al. (2009).
[b] Thomson et al. (2011).
[c] Riahi et al. (2011).
[d] IPCC (2014a).
[e] Chang, Guo, and Xia (2012).

slows (15–60 percent). The frequency of extreme El Niño events increases (W. J. Cai et al., 2014). The northern hemisphere storm tracks shift north (Chang et al., 2012).

Globally applied biogeography models (e.g., DGVMs) driven by the ESM outputs suggest a displacement of global vegetation types (Yu et al., 2014), notably poleward movement of forests in the northern and southern hemispheres. CO_2 fertilization causes increases in tree and shrub cover globally, along with greater vegetation productivity. These projections are generally consistent with observed trends in global vegetation over the last 30 years (see chapter 5). The terrestrial biosphere is projected to change from a carbon sink to a carbon source because of more disturbances (e.g., fire) and effects of warming on rates of decomposition (Muller et al., 2016).

Environmental change of the magnitude suggested by the BAU scenario would challenge the adaptive capacity of even the most advanced technological civilization. Impacts in the least developed countries, with limited adaptation capacity, would be severe. Most of the trends indicated in chapters 4 and 5 would likely continue. As noted, the incidence of extreme weather events would increase, along with changes in the mean values. Agricultural production, which would need to increase substantially to feed a larger global population, would be impacted by erratic weather and changing climate. Marine fish and shellfish production might decrease in response to acidification, stratification, and warming. Sea level rise would force global retreat from coastlines (Hauer, Evans, and Mishra, 2016), potentially affecting nearly a billion people by 2060 (Neumann, Vafeidis, Zimmermann, and Nicholls, 2015).

Greenhouse Gas Stabilization Scenario (RCP 4.5)

The key feature of this scenario (table 6.1) is that climate policies (e.g., carbon taxes and reductions in deforestation) are introduced to reduce greenhouse gas emissions. Technological advances such as CO_2 capture and geological storage are assumed, as well as expansion of forests. CO_2 emissions climb to 11.3 GtC per year around 2050 and then decline to 3.6 GtC per year by 2100. The CO_2 concentration hits 525 ppm and stabilizes by 2100.

Midrange ESM projections for the stabilization scenario in 2100 suggest an increase in global mean temperature of 1.8°C and sea level rise of 0.17 m. The kinds of changes in the storm tracks found in the BAU scenario are also seen in the stabilization scenario model runs, but are slower to appear (Chang et al., 2012). Ocean pH decreases and AMOC slows, but there is no suggestion of abrupt AMOC collapse (Weaver et al., 2012). The projected impacts on ecosystems and human welfare are less than for the RCP 8.5 scenario but would still strain adaptive capacity.

Greenhouse Gas Reduction Scenario (RCP 2.6)

The essential feature of the reduction scenario is that a much larger commitment is made to reduce emissions and increase sequestration than was the case in the stabilization scenario, and by 2100 the CO_2 concentration has begun falling. The CO_2 concentration peaks around 2050 at 450 ppm and declines to 425 ppm by 2100. High enough carbon taxes are imposed to force down CO_2 emissions, and bioenergy (crop residues and municipal waste) with carbon capture and burial makes a large contribution to total energy supply and to removing carbon from the atmosphere. Note that these are rather heroic assumptions relative to what we know at present about CO_2 removal from the atmosphere (P. Smith et al., 2016). Massive investments in nuclear or wind energy, or both, are also assumed. James Hansen and colleagues (2013) laid out a similar scenario that aimed for 350 ppm by 2100. That scenario required a 6 percent per year reduction in fossil fuel emissions beginning in 2013 and large carbon sinks from reforestation.

The impacts of the reduction scenario are generally only a bit greater than the changes already observed in recent decades. In addition to the projected impacts from current ESMs (table 6.1), the increased CO_2 would likely increase global mean temperature an additional few tenths of a °C by reducing stomatal conductance (Randerson et al., 2015). Arctic Ocean ice in September would remain about half of the late twentieth century values (IPCC, 2014b).

The results of the 2014 IPCC assessment are well described and summarized elsewhere (IPCC, 2014a). There are separate volumes on climate change itself (IPCC, 2014b), the related impacts (IPCC, 2014c), and possible mitigation strategies, including geoengineering options (IPCC,

2014d). Besides the IPCC global assessments of potential climate change and its impacts, there are various assessments at the national (NCA, 2014a), regional (NCA, 2014b), and state (Miles, Elsner, Littell, Binder, and Lettenmaier, 2010) levels. Assessments over smaller domains commonly use climate projections that have been "downscaled" from the coarse spatial resolution of the ESMs to something on the order of grid cells measuring 1–25 kilometers on a side (Abatzoglou and Brown, 2011).

LONGER TIME HORIZONS

Scenarios to 2300

Because fossil fuel emissions continue after 2100 in three of the four RCP scenarios, and slow climate feedbacks will take centuries to play out, there is considerable interest in how climate will change in the longer term. IPCC climate modelers were therefore also provided with extensions of the RCPs out to 2300 (i.e., extended concentration pathways). These concentration projections were not produced by IAMs because of the high uncertainty in economic forecasts (e.g., their sensitivity over long intervals to assumptions about the discount rate; see Ackerman et al., 2009). In some cases, these scenarios assume withdrawals of CO_2 from the atmosphere and hence reductions in concentration after 2100. In all cases, it is assumed that fossil fuel emissions become very small by 2300 because of mitigation or because nearly all accessible fossil fuel has been burned. CO_2 concentration rises to about 2,000 ppm in the BAU scenario by 2300 and drops to 350 ppm in the reduction scenario.

To investigate Earth system responses out to 2300, ESMs of intermediate complexity are usually employed (Randerson et al., 2015). In the time frame of the extended concentration pathways (out to 2300), warming continues to increase in most scenarios even if the greenhouse gas concentrations stabilize or begin to fall. The increase in the global mean temperature for the BAU scenario is approximately 8°C. Precipitation continues to decrease in places, "exceeding a factor of two over parts of Australia, the Mediterranean, southern Africa and the Amazon, and . . . exceeding a factor of three over parts of central America and North Africa" (Tokarska, Gillett, Weaver, Arora, and Eby, 2016, p. 854). The warming response is associated with slow changes in the cryosphere (e.g., the melting of the Greenland ice cap and the general decrease in the amount of highly reflective snow cover). These changes provide a continuing positive feedback to warming. Projected sea level rise ranges from 0.5 to 3.5 meters. Vegetation likewise continues changing even if CO_2 is stabilized (Salzmann, Haywood, and Lunt, 2009). The ocean continues as a carbon sink, but its carbon uptake efficiency per unit CO_2 increase diminishes because of reduced AMOC and warmer water temperatures (warmer water holds less CO_2).

Even given rapid rates of change in the technosphere and in human capacity for adaptation, it is difficult to imagine a world in 2300 that is 8°C (or more) warmer than the present climate. Significant areas would become largely uninhabitable because air temperatures greater than 35°C are metabolically unsustainable for humans. At a 12°C increase "such regions would spread to encompass the majority of the human population as currently distributed" (Sherwood and Huber, 2010). A sea level rise of 3 meters or more would severely impact many coastal cities globally (SLRMap, 2017).

Scenarios to 10,000 Years

The slug of CO_2 ejected into the atmosphere from combustion of fossil fuels will likely be around for a geologically significant period (Archer, 2005). Indeed, it is beginning to look like the climate system may miss the next orbital forcing induced increase in glaciation, expected sometime between now (Ruddiman et al., 2016) and 50,000 years from now (Berger, Loutre, and Crucifix, 2003). A recent effort was made to evaluate sea level change out 10,000 years (P. U. Clark et al., 2016): the estimates ranged between 20 and 50 meters. By comparison, sea level rose more than 100 meters in the previous 10,000 years as the Earth system came out of the last glacial period.

Assuming CO_2 concentration remains on the order of 500–700 ppm for 10,000 years (P. U. Clark et al. 2016), Earth will return at least to the warmer, wetter conditions of the middle Pliocene epoch (2.6–3.6 Mya [million years ago]). At that time, the northern tree line was shifted to higher latitudes and tropical savannahs and woodlands occupied what are now deserts in many areas (Salzman et al., 2009). A possible analogue to an even warmer climate than the Pliocene (Pagani et al., 2009) is the mid-Miocene climate optimum (15–17 Mya), when global mean temperature was approximately 4°C –5°C warmer than today. At that time, cool temperate forests were found at high northern latitudes (Pound, Haywood, Salzmann, and Riding, 2012).

To find anything similar to the 8°C increase in mean annual temperature projected in the BAU scenario, we must go back to the early Cenozoic era (approximately 50 Mya). Recall from chapter 3 that Earth transitioned during the Cenozoic era from a greenhouse world with high CO_2 and limited glaciations in Antarctica, to an icehouse world with lower CO_2 and ice caps over Greenland and Antarctica (Frakes, Francis, and Syktus, 1992). In the greenhouse conditions of the early Cenozoic, forests were present on Antarctica. Given CO_2 above 600 ppm and time to melt the ice caps, Earth could return to that greenhouse state (Haywood et al., 2011).

THE ICARUS SCENARIO

We must also face the possibility that human-driven increases in greenhouse gas concentrations will push the Earth system over a threshold in which strong positive feedbacks to warming are engaged and, over a multimillennial time frame, environmental changes lead to a mass extinction event. Here, we will call this the Icarus scenario. Icarus is the figure from Greek mythology who constructed wings of feathers and wax, then flew too close to the sun. The wax melted, the wings disintegrated, and Icarus crashed to Earth. It is an apt metaphor for an Earth system scenario in which humans build a technosphere that unintentionally disrupts the climate system and the biosphere to such a degree that Earth no longer supports an advanced technological civilization.

As we saw in chapter 2, the paleorecord documents compelling case studies of greenhouse gas–driven warming leading to ocean stratification and large perturbations of the global biogeochemical cycles, and, eventually, to extinction events (Norris, Turner, Hull, and Ridgwell, 2013). The current state-of-the-art in Earth system modeling is not adequate to simulate this kind of future because of limitations in scientific understanding of the processes, the complexity of their potential interactions, and the extended time spans over which they may play out. Further, because the current rapid rate of increase in greenhouse gas concentrations has no precedent in geological history, we are already in no-analogue territory. That makes it even more difficult to simulate the long-term future.

Earth system scientists have identified several potential tipping points or elements in the Earth system that could strongly respond to climate change and amplify it (Y. Y. Cai, Lenton, and Lontzek, 2016; Lenton et al., 2008). The tipping point concept holds that there are threshold values (forcings) for these elements or processes in the Earth system beyond which the process responds nonlinearly or undergoes a phase change. Beyond the tipping point, the process rate is controlled by internal dynamics more than by the climate; i.e., it becomes self-perpetuating. Some of the tipping elements that will amplify climate warming include the following.

1. *Melting of the Greenland ice sheet.* As melting proceeds and the elevation of the ice surface descends (it is 2–3 kilometers deep), the surface ice is exposed to an ever-warmer climate. The threshold change in global mean temperature beyond which the Greenland ice sheet will be committed to an ice-free state was recently estimated at 1.6°C (A. Robinson, Calov, and Ganopolski, 2012).

2. *Disintegration of the West Antarctic ice sheet.* Either floating ice shelves or ones grounded below sea level buttress the upstream grounded ice around Antarctica. These buttress formations have begun melting, and once removed, the rate of ice flow to the ocean will accelerate (Fürst et al., 2016). As more land and water are exposed, the snow/ice albedo feedback strengthens.

3. *Collapse of the THC (AMOC and Antarctic overturning circulation).* The THC is responsible for sinking water in the North Atlantic and rising cold water around Antarctica, both cooling influences on the climate. As the climate warms, the THC appears to be slowing. Hawkins and colleagues (2011) suggest that the THC is bi-stable, i.e., tending to be either on or off.

4. *Dieback of the Amazon rain forest.* Since a significant proportion (20–35 percent) of rainfall in the Amazon Basin is recycled, loss of large areas of forest could alter the regional climate such that it no longer supports rain forests. The combination of declining rainfall in the southern Amazon, and deforestation, could thus induce positive feedbacks to

forest loss mediated by the hydrologic cycle (Malhi et al., 2009). Forest loss is a source of additional CO_2 to the atmosphere.

The Icarus scenario assumes the BAU scenario out to 2300. The warming associated with 1,000–2,000 ppm CO_2 pushes the Earth system through one or more of these tipping points. The additional warming from tipping point elements is assumed to induce a further positive feedback—melting of marine methane hydrates (Archer, Buffett, and Brovkin, 2009). This would be a relatively slow process, and methane has a short atmospheric lifetime. But it oxidizes to CO_2 and that CO_2 source maintains upward pressure on greenhouse gas concentrations for centuries, which offsets the silicate weathering negative feedback to CO_2 increase and warming (see chapter 2). Over millennia of high CO_2 conditions, ocean stratification strengthens, much of the ocean becomes anoxic, and the microbial metabolism in the ocean shifts toward green sulfate-reducing bacteria and hydrogen sulfide production (Kump, Pavlov, and Arthur, 2005). Hydrogen sulfide is toxic to most life forms, and depletes stratospheric ozone (hence increasing mutagenic ultraviolet radiation). A major extinction event occurs.

As noted, because of its complexity and scale, and many unknowns (e.g., the size of the marine methane hydrate reservoir), Earth system scientists are not able to rigorously quantify the tipping points (Kriegler, Hall, Held, Dawson, and Schellnhuber, 2009) and cannot mechanistically simulate the Icarus scenario. It serves mostly as a downside end point of the continuum of Earth system scenarios. One thing we do know is that something like it has previously occurred: the end-Permian mass extinction.

THE EQUILIBRATION SCENARIO

In our idealized Anthropocene narrative, outlined in chapter 1, the Great Acceleration is followed by the Great Transition. The IPCC greenhouse gas reduction scenario is a model out to the year 2100 for this transition away from exponential growth. However, an upbeat vision for the planet beyond 2100 is also desirable. It might be called the "equilibration scenario." The term *equilibration* has distinct meanings in the fields of physics, biology, and psychology—generally referring to a process by which two forces are balanced (e.g., the exchange of energy between two adjacent objects that start at different temperatures). Regarding the long-term future of our planet, the equilibration scenario evokes a sustainable integration of the technosphere with the rest of the Earth system (see "Long-Term Effects of Ecological Modernization," in chapter 7). Certainly, in creating a range of Earth system scenarios, we can aspire to building a sustainable high-technology global civilization (Grinspoon, 2016; Raskin, 2016) that might even survive over geological time scales (M. Williams et al., 2015). Arguably, a positive vision of the future will make that vision more likely to come about.

ESSENTIAL POINTS

We can't know Earth's environmental future, but the Earth system science community is gaining an increasingly sophisticated capacity to simulate that future based on mechanistic mathematical models. These coupled models consider the biosphere, atmosphere, hydrosphere, and cryosphere, as well as human factors such as demography, energy systems, and economic development. A widening array of Earth system observations (which we will discuss in chapter 9) allow for design, calibration, and testing of the models. Depending on the success of efforts to limit greenhouse gas emissions, the magnitude of projected changes in global mean temperature, sea level rise, atmospheric circulation, and ocean circulation vary widely. However, the BAU scenario for greenhouse gas concentrations will clearly cause profound, geological-scale changes in the global environment.

IMPLICATIONS

1. Fossil fuel emissions over the next 100 years will determine the magnitude of a change in global climate that will likely dominate the Earth system for thousands of years. The climate change will drive a rise in sea level, shifts in the jet stream (causing changes in the distribution of

precipitation and in extreme events), and loss of mountain snow and glaciers (that buffer summer stream flow).

2. In the intermediately optimistic IPCC greenhouse gas stabilization scenario (RCP 4.5), a global demographic transition is completed by around 2100 and carbon emissions decline substantially from the present. However, the Earth system may nevertheless transition in the following centuries to a greenhouse planet climate similar to that of the Cenozoic era around 30–50 Mya.

3. Given the momentum in anthropogenic greenhouse gas emissions and the long residence time of atmospheric CO_2, it is likely the next ice age will be deferred.

4. In the worst-case Icarus scenario, rising greenhouse gas concentrations cause strong climate system feedback mechanisms to be engaged, which profoundly transform the global biogeochemical cycles, eventually (over thousands of years) inducing a massive extinction event. This scenario cannot be ruled out.

FURTHER READING

Hansen, J., Kharecha, P., Sato, M., Masson-Delmotte, V., Ackerman, F., Beerling, D. J., . . . Zachos, J. C. (2013). Assessing "dangerous climate change": Required reduction of carbon emissions to protect young people, future generations and Nature. *Plos One, 8*, 26.

7

GLOBALIZATION AND ECOLOGICAL MODERNIZATION

Changes in human societies over the past 12,000 years can be understood as constituting a single complicated earth-wide event of spiraling globalization.

—C. Chase-Dunn and B. Lerro (2013)

SOCIOECONOMIC MODERNIZATION

A foundational concept used by historians of Western civilization to capture the dynamics of the last couple of centuries is "modernization." The concept is interpreted in various ways by different disciplines (and factions within disciplines), but here we will consider it as a form of socioeconomic development broadly associated with the process by which Western agrarian societies were transformed into industrial societies. Modernization meant moving beyond the religious and political dogmas of the medieval era toward greater religious freedom, political freedom, and the advance of science and technology. Paradoxically, it fostered individualism and emphasized humans as isolated economic agents (Polanyi, 1944), but also favored the elaboration of ever larger and more hierarchically structured societies (Giddens, 1991). Modernization has been driven especially by expansion of the scientific worldview, by capitalism, and by energy derived from combustion of fossil fuels.

The Western modernization narrative mostly focuses on social, intellectual, and technological changes (progress), with the environment largely in the background. If anything, modernization has been viewed as emancipation of society from Nature. However, as we have seen in chapters 4 and 5, the technological advances associated with modernization and their pervasive application have degraded local environments and begun to alter the global environment. Beginning in the 1960s, a counter-narrative to Western modernization began to coalesce around concerns about the dark side of the process. Regarding the environment, that meant emphasizing the role of modernization in creating a world vulnerable to nuclear holocaust, widespread environmental degradation, and climate change (Beck, 1992, 1999). Regarding societal development, it meant acknowledging widespread economic inequality and social injustice (Malm and Hornborg, 2014). Besides generating improved standards of living and wealth, modernization was recognized as creating risks to individuals and societies. Sociologists suggested the need for a critical examination of modernization— and a "second modernization" to reform it (Beck, Giddden, and Lash, 1994).

The Western model of modernization is by no means a template for societal dynamics everywhere (Beck and Grande, 2010). Modernization theorists in the early Cold War era suggested that "poorer nations are poor because they lack adequate capital, technology, and modern social organization and values" (Hite, Roberts, and Chorev, 2015, p. 9). There is certainly more to it than that. Value judgments regarding Enlightenment ideals, individualism, education, capitalism, and representative government are highly divergent, as are endowments of natural resources. Also, the earliest developing countries may exploit the less developed countries and impede their sociocultural evolution (Chase-Dunn and Lerro, 2013). In terms of impact on the environment, however, the critical point is that technological advances and the benefits of fossil fuel combustion have been adopted widely in the process of economic development. Thus, the local, regional, and global environmental impacts of widespread socioeconomic development now require a significant measure of rethinking about the process itself and its consequences.

The human dimension of Earth system science includes the work of environmental sociologists, political scientists, and economists who study socioeconomic development and environmental governance at all scales. Unlike the geophysical and biological domains of Earth system science, the research and theorizing of social scientists who study the Earth system can take on normative implications (i.e., what ought to be done).

RESPONSES TO MODERNIZATION

Modernization spawned the technosphere. Now, increasing recognition of the threat that the technosphere poses to the biosphere (i.e., the planetary life support system) has engendered several sorts of responses. One is a call for demodernization or degrowth; i.e., an attempt to simplify our technologies and ways of living to minimize environmental impacts. A second response is associated with the postmodernism movement, which questions the dominant Western paradigm of endless progress, but avoids giving special status to any viewpoint, including the ecological worldview. The effect of postmodern thinking with respect to the environment is essentially business-as-usual (BAU) policies. Alternatively, environmental sociologists have identified the path of "ecological modernization." Here, the environment is given equal import with the economy, and humans take responsibility for managing their impact on the environment at all scales. Let's consider each response separately.

Demodernization

The response of demodernization is based on a retreat from complex technology. It advocates smaller, more self-sufficient social units. The philosophical underpinnings of demodernization come from the Deep Ecology movement (Devall and Sessions, 1985). Deep ecologists question whether humanity has the right to appropriate as large a proportion of Earth's resources as we currently consume. A key principle of Deep Ecology is that the intrinsic worth of other species on the planet is the same as the worth of humans. The degrowth movement (Fournier, 2008) similarly challenges the unbounded growth in resource use, especially in the case of energy.

The revolutionary nature of demodernization/degrowth is attractive because, as we have seen (chapters 4, 5, and 6), there is considerable urgency about reducing fossil fuel emissions, deforestation, and other anthropogenic impacts on the global environment. Inherent in this response is a questioning of the market-based economic system that underlies the current global patterns in production and consumption. The title of a popular environmental science book—*This Changes Everything: Capitalism vs. the Climate* (2014)—is a call-to-arms for revolutionary change in the economic system. Author Naomi Klein, a journalist, questions the basic premises of capitalism and advocates mass social movements, as in the late 1960s United States, to change it. In the 1960s, the issues of racism, the Vietnam War, political corruption, and a deteriorating environment inspired large-scale demonstrations, and subsequently progressive legislative actions. A revolutionary movement to address global environmental change based on moral fervor is now espoused by philosophers as well as religious figures (Moore and Nelson, 2011; Schiermeier, 2015). The difficulty, of course, is that a rapid retreat from technology and capitalism might well end with societal collapse in both the developed and developing world.

A common counterargument to the proposition that capitalism is inherently destructive to the environment is the observation that when economic development reaches a certain level, environmental quality begins to improve. The environmental Kuznets curve (figure 7.1) indicates increasing pollution in the early state of economic development as the scale of industry increases, a maximum in a fully industrial economy, but then a falling off in a postindustrial (more service-oriented) economy. The environmental Kuznets curve implies that investments will be made to reduce pollution once economic growth provides reliable food, shelter, and security. As discussed below, economic development provides the financial capital to support those investments.

Environmentally oriented economists largely reject the "environmental restoration through wealth creation" proposition (Suri and Chapman,

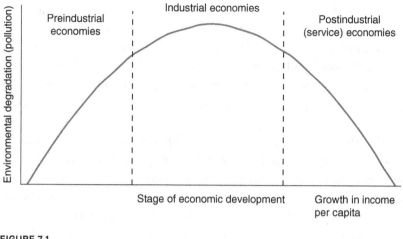

FIGURE 7.1

The environmental Kuznets curve.

1998). However, national-level case studies in recently developed and currently developing countries lend some support to this model (Azam and Khan, 2016; Pandit and Paudel, 2016). The most recent national-level survey of trends in forest area and economic development found an increase in forest area associated with an increase in gross domestic product (GDP) per capita (Morales-Hidalgo, Oswalt, and Somanathan, 2015). Certainly, many developing countries are still in the upward trending part of the curve. A significant constraint in rounding the curve is that a regulatory approach to environmental quality is required, but governance capacity is often limited in developing countries (Huber, 2008).

The Postmodernism Response

The postmodernism path is in some ways difficult to even think about because the postmodern critique of modernism questions any sort of reference narrative or worldview (Harvey, 1989). Most fundamentally, it questions the "grand narrative" of world history based on the rise (progress) of

Western civilization. One beneficial result of the postmodernism movement has been an emphasis on multiculturalism, especially in academic environments. This respect for diversity creates a sense of global community and hence a collective response to the challenges of global environmental change. The postmodern perspective is also helpful in questioning the dominant Western cultural narrative of technology-based conquest of Nature.

The limitations of the postmodern mindset are its divisiveness and reluctance to give the scientific worldview a special status. The postmodern perspective supports questioning authority and liberation from dominant belief systems, which leads to fragmentation. If the threat of climate change were at last to be recognized globally and began to serve as a universal reference point for important policy decisions, a postmodernist would evince suspicion of this preeminent viewpoint. The problem is that, indeed, the stark observations of human-induced global environmental change (outlined earlier in chapters 4 and 5) and the projections for BAU-induced changes in the coming centuries (chapter 6) do require special societal attention. The threat to the biosphere as the planetary life support system does, in fact, give everyone on the planet a common reference point.

Ecological Modernization

Ecological modernization recognizes the environmental threat posed by BAU. However, it assumes that technology will evolve sufficiently to avoid the BAU scenario and that global environmental problems can be addressed within the framework of the current, dominant, socioeconomic system (Spaargaen, Mol, and Buttel, 2000). Thus, it follows a reformist rather than a revolutionary model of societal change. The fundamental transformation espoused by ecological modernization is to bring ecological considerations to the same level of importance as economic considerations in all societal deliberations. It is a path to the Great Transition.

Before examining ecological modernization theory, let's visit globalization as a force that is spreading modernism, and hence magnifying the threat of global environmental change.

GLOBALIZATION AND ANTIGLOBALIZATION

Globalization as a concept refers to both the compression of the world and the intensification of consciousness of the world as a whole.

—R. Robertson (1992)

The Two Faces of Globalization

The term *globalization* refers in a general way to the increasing interconnectedness of individuals and social groups everywhere on the planet, and the increasing inability of any particular social group to isolate itself from outside influences, especially economic ones. Broadly considered, it begins with the fifteenth century exploration out of Portugal and Spain (Chase-Dunn and Lerro, 2013). Teilhard de Chardin, as early as the 1920s, referred to a global "compression" induced by increasing cross-border flows of people, products, and information. Nearly 100 years later, humanity is riding the crest of the Great Acceleration of technosphere expansion (Hibbard et al., 2007) and, indeed, there is nowhere to hide.

Globalization contributes to the degradation of the global environment by spreading the Western model of industrial development, characterized by the commodification of natural resources, consumerism, and burning of fossil fuels. However, it also stimulates the kind of communications network and intensive international interactions that are needed to address the wide array of emerging global environmental change issues (box 7.1). Additionally, globalization facilitates the rapid transfer of environmentally friendly technologies from the developed to the developing world.

The contemporary globalization process can be broadly broken out into economic, political, and cultural dimensions. *Economic globalization* is its most familiar aspect, and refers to the penetration of the mostly capitalism-based market system to all corners of the world. Institutions such as the World Trade Organization (WTO) and the International Monetary Fund (IMF) work to minimize barriers to international trade and promote economic development (see chapter 8). WTO trade agreements

BOX 7.1

The Globalization of Sociocultural Evolution

Sociocultural evolution refers to "processes by which variations in cultural information are produced, transmitted, selected for, change in frequency, and accumulate across generations and across sociocultural systems" (Ellis, 2015, p. 293). Early human societies were groups of hunter-gatherers, but the scale of sociocultural systems has shifted (i.e., evolved) over historical time to empires, nation-states, and civilizations.

Much of the culture within a society is generated internally, but cultural replicators (memes) from outside can also be introduced and proliferate. Until recently, the exposure to external memes was limited in traditional societies. However, we now have a global telecommunications infrastructure (including the Internet) by which memes are rapidly spread. The number of active mobile communication devices, many of them smart phones, now exceeds the number of people on the planet (Boren, 2014). Thus, we are entering an era of sociocultural evolution at the global scale. Innovative memes (or memeplexes) such as nationalism, capitalism, and science have recently achieved a global distribution.

Addressing the range of global environmental change issues (chapters 4 and 5) will require that concern about environmental change become a core element of the emerging global cultural conversation. Unlike biological evolution, sociocultural evolution involves a significant element of intentionality (Chase-Dunn and Lerro, 2013). That puts a burden on scientists, educators, and media specialists to make the global environmental change memes heard widely. Shifts in cultural norms can be instigated and propagated by information exchange.

suppress protectionism-based economic strategies, epitomized by tariffs on imported goods. The consequence has been offshoring of manufacturing from developed countries to benefit from cheaper labor and, usually, more limited regulations (notably environmental regulations) elsewhere. The cost to the developed countries is loss of low-skilled jobs, whereas the benefit is lower prices on manufactured goods.

A notable environmental downside of economic globalization has been displacement of pollution from manufacturing facilities in developed countries to those in developing countries. As manufacturing facilities moved offshore, the associated waste products accompanied them. Environmental regulations tend to be weaker in developing countries; hence the sources of pollution may be greater and persist longer. This environmental cost shifting raises ethical issues in the context of rights-based versus wealth-based distribution of pollution (J. K. Boyce, 2004). The transfer of manufacturing to developing countries has pushed the total global emissions ever higher because of increased emissions associated with expanded transportation of raw materials, manufactured goods, and people. More generally, globalization has increased income and consumption, thus increasing the ecological footprint of humanity.

At the global scale, a positive social consequence of economic globalization has been to lift hundreds of millions of people out of poverty (Economist, 2016a). Much of that poverty reduction was in China. People in extreme poverty are susceptible to degrading their local environment; hence reduction in their numbers is potentially a benefit to the environment. There has also been a tendency for environmental standards in developing countries to improve as a result of globalization, because (1) environmental quality receives more attention locally as living standards improve (figure 7.1), and (2) consumers in developed countries begin to seek out products from developing countries that have been certified as sustainable (see chapter 8).

Political globalization refers to the increasing level of interaction and negotiation among nation-states regarding military, economic, and, most recently, environmental issues. The United Nations has become a forum for attempting to resolve military conflicts. The WTO, IMF, and World Bank are some of the international institutions that organize global economic governance. As will be discussed in chapter 8, global environmental governance has been more ad hoc than economic governance, and needs greater institutional identity and stability (Biermann and Bauer, 2005).

Cultural globalization is associated with the rapid spread of mostly Western ideas, beliefs, and values to geographic areas formerly dominated by local traditional cultures. English has tended to become the language of the global commerce and the global science communities. The Hollywood entertainment industry now makes more money internationally than it does in the United States. Tourism is said to be the world's largest industry and accounts for about 10 percent of global domestic product. The Internet, with its promiscuous fountain of words, ideas, and images, is of course greatly increasing the rate of cultural globalization. The cost to the environment is more consumerism; the benefit is a growing awareness of global environmental change issues.

Cultural globalization is provoking fundamentalist responses, but it is also precipitating the coalescence of a global society. Again, the latter could be considered beneficial from an environmental perspective in the sense that addressing global environmental change issues will require a unified global response. Earlier in this book we noted that in the environmental sociology literature, there is widespread interest in the concept of the "global risk society" (Beck, 1999). The idea is that the environmental crises generated by the modern industrial era have become a major factor in social organization. Environmental problems cannot be escaped by anyone on the planet; hence a global society is forming to confront those problems (Dinar, 2011).

The convergence in socioeconomic systems among the nations of the Earth on nationalism, democracy, and capitalism has been characterized as the "end of history" (Fukuyama, 1992). This formulation grew out of the end of the Cold War, when communism as an economic system was widely recognized as unable to keep pace with capitalism. This notion is, of course, overblown and notably ignores global environmental change. Humanity as a collective whole faces a multigenerational challenge to mitigate and adapt to anthropogenic environmental change (Ehrlich and Ehrlich, 2012).

The advance of science is a fundamental component of cultural globalization, as a contributor toward the spread of modernism. In the scientific worldview, the individual mind is freed of religious and political dogma and able to set out on its own to discover a quasi-objective reality. Within the global scientific community, the mind of an individual scientist, or the collaborative effort of a group of scientists, assimilates the history of ideas and observations related to a discipline, generates new hypotheses about causal relationships among the objects and processes of study, and then goes about testing them. Results of these speculations, experimental tests, and observations are packaged into scientific papers, subjected to peer review, and communicated to the scientific community in journals and books. Upon publication, the ideas receive further review by the scientific community and the public, and are selected, in terms of cultural evolution, by being referred to in subsequent publications by others. Thus, a model of how the world is structured and how it functions is being continuously built up and revised. In *Consilience*, Harvard biologist E. O. Wilson makes the case that the scientific worldview will ultimately colonize almost all realms of knowledge (E. O. Wilson, 1999).

Antiglobalization

Globalization is, of course, not always to everyone's benefit. Large protests have disrupted several meetings of the WTO, the World Bank, and related bodies, notably the 1999 WTO meeting in Seattle. The protesters—included unionists, environmentalists, and anarchists—aimed to draw attention to the nontransparent nature of WTO negotiations and the way associated agreements favored corporations over labor rights, the environment, indigenous people, and less developed nations. There was fear that these agreements promote a "race to the bottom" in terms of environmental protection and social values.

From an environmental perspective, the downside of international free trade agreements is that legally binding decisions, having significant environmental implications, are made on mostly economic grounds. In contrast, environmentalists favor trade restrictions on environmental grounds (e.g., against shrimp imports from countries that do not use sea turtle protection devices required by U.S. domestic producers). Environmental issues are gaining increasing prominence in trade negotiations (SEDAC, 2016), e.g., the WTO updates of the General Agreement on Tariffs and Trade (WTO, 2017). However, the prospects for expanding existing trade agreements and developing new multilateral trade agreements are diminishing in the face of sentiment against economic globalization.

The political push-back against globalization in the highly developed countries of Britain (the vote to exit the European Union) and the United States (the 2016 election results) is based on the perception that the economic benefits of globalization flow mainly to the relatively wealthy segment of the population, whereas the costs—in terms of loss of domestic jobs—are borne by everyone else. More broadly, however, antiglobalization includes a negative sentiment toward international entanglements and commitments. The national commitments to the 2015 Paris Agreement on climate change could translate into policies such as carbon taxes and subsidies for renewable energy, which would cut into demand for traditional coal and oil products. Large proportions of the population see the costs of mitigating climate change, but not the benefits.

Mol (2000) refers to the "apocalypse blindness" of the old, mostly national-level, institutions; specifically, their mandates generally do not extend to addressing emerging global-scale risks. Unfortunately, antiglobalization fervor may slow the evolution of the new national-level and global-level institutions (see chapter 8) needed to mitigate and adapt to technosphere-induced changes in the Earth system.

Indigenous People's Rights

Economic globalization tends to put pressure on unexploited natural resources everywhere on the planet. Land grabs against indigenous people in developing countries are thus common. Often the land is used sustainably before it is appropriated, and is converted to production of export commodities afterward, including tropical hardwoods, beef, palm oil, and soybeans.

Rather than surveying the many instances of this process globally, let's focus on one particularly fascinating case study. It relates to construction of the Belo Monte hydropower dam on the Upper Xingu River in the Amazon River Basin (Economist, 2013). The land to be flooded by the dam had already been designated as a national park and reservation for indigenous people. Nevertheless, after considerable controversy the Brazilian government approved a plan for construction of a dam, which is now operating. The dam was promoted as a needed source or renewable energy for regional development, and its construction costs were subsidized by the Brazilian government.

Indigenous people living along the banks of the Xingu River resisted construction approval for years, and some accommodations were made by changing the size of the reservoir and making investments in social programs for the displaced residents (20,000–40,000 people). The construction of the dam brought thousands of people to the area who have imposed widespread land use change. The environmental costs included the likely extinction of several endemic fish species.

This conflict has many of the hallmarks of globalization, notably the compression of multiple previously isolated social entities. Economic globalization is relevant in relation to expanding Brazilian exports (e.g., aluminum) produced with power from the hydroelectric dam. Cultural globalization is evident in the global media coverage of well-known people, including Sting, James Cameron, and Sigourney Weaver, who appeared in support of the indigenous people. An ironic feature of globalization is that even if a societal or cultural group is trying to resist outside forces, it may well use some aspect of globalization (in this case the global media) in its efforts.

The Amazon Basin in Brazil has significant potential for hydroelectric energy production, and planners have already mapped out a network of dams (more than 150) for the region. Several dams are proposed upriver from Belo Monte to help maintain flows to the hydroelectric turbines during the dry season. Along the Tapajos River to the east of the Xingu River, the history of the Belo Monte Dam is being replayed (Economist, 2016b). The federal government approved the first of a series of dams on the river, but recently reversed course based on issues with indigenous people. Whether this change is part of the Great Transition, or just a bump in the road of the Great Acceleration, is not clear.

ECOLOGICAL MODERNIZATION THEORY

Ecological modernization (EM) is a process by which societies address environmental change issues. Ecological modernization theory (EMT) is a theory of social change, i.e., an effort to explain why and how EM occurs. EMT is based on the assumptions that (1) science and technology can provide the goods and services that we (the global billions) aspire to, without degrading the environment; and that (2) the changes that need to be made to restore and maintain environmental quality are economically and politically feasible. EMT developed in Western Europe and is a product of reflexive modernization that builds on the western model of modernization (Spaargaen et al., 2000). The degree to which it applies to environmental management in non-European countries, and at the global scale, is under debate.

EMT accepts that modernization and globalization have induced a contemporary environmental crisis, but it anticipates that technology will increasingly become environmentally friendly and provide solutions to emerging environmental problems. Atmospheric scientists discovered that smokestack emissions of sulfur and nitrogen were causing aquatic acidification, so engineers invented smokestack scrubbers that remove the sulfur and nitrogen. Now, Earth system scientists have discovered that carbon dioxide (CO_2) from coal-burning power plants is accumulating in the atmosphere and causing global warming, so engineers will design and build solar and wind turbine facilities to generate power from renewable sources. There is a strong reliance in EMT on the worldview of the scientist and the engineer, and on the effectiveness of governmental policy to ameliorate environmental problems.

At the level of individuals, EMT means reducing personal consumption (e.g., biking, recycling), the greening of consumer choices (Spaargaen and Mol, 2008), a willingness to pay taxes for improving environmental quality, and the greening of electoral politics. Informed consumers purchase environmentally friendly products. Informed voters elect environmentally friendly politicians. Citizens are assumed to engage in civil society in a way that influences the evolution of environmental policy. The old divisions in society among socioeconomic classes are overridden by a common exposure to risk from environmental change (Beck, 1992).

At the level of social organizations, EM means a rich array of nongovernmental organizations (NGOs) that address environmental change issues. These would include organizations such as (1) the Forest Stewardship Council, which certifies wood products as produced sustainably; (2) the

Nature Conservancy, which purchases and manages small and large tracks of land for conservation purposes; (3) the Natural Resources Defense Council, which lobbies and litigates for environmental causes; and (4) Greenpeace, which undertakes direct action against companies that degrade the environment. NGOs work collaboratively with governmental agencies.

At the level of nation-states, EM involves building appropriate (strong but not overbearing) regulatory frameworks, and using the regulated market to incentivize environmentally friendly development. The state is seen as sensitive to bottom-up pressure for environmental management reforms. EM is favored by "political modernization," meaning governments that thrive on open debate (Leroy and van Tatenhove, 2000). To address regional and global-scale environmental change issues, EM requires some surrender of national sovereignty, e.g., through participation in international environmental agreements.

At the regional to global scales, EM involves interstate connections to address environmental issues by way of diverse international nongovernmental organizations (INGOs). EM downplays the importance of interstate conflicts and points toward collective commitments to addressing the risks of regional and global environmental change. Consistent with that approach, the European Union and NAFTA (the North American Free Trade Agreement), while originally created in part to promote trade, now address significant environmental issues. The expanding array of global treaties on the environment, notably the Paris Agreement on climate change, indicates the growing potential for international collaboration on global environmental change issues. INGOs become a source of activism about global change issues, not just a response to them (see chapter 8).

As noted, EMT is subject to a variety of criticisms and corresponding defenses (Fisher and Freudenburg, 2001; C. L. Smith, Lopes, and Carrejo, 2011). Indeed, the following issues must be faced in the ongoing development of EMT.

1. The current socioeconomic paradigm of capitalism inherently relies on endless growth, which is incompatible with biophysical limits to the expansion of the technosphere (R. Smith, 2015).

Economists increasingly question this view of capitalism (Saunders, 2016). As standards of living improve, rates of economic growth tend to decline. The build-out of modernity (i.e., the infrastructure of housing, transportation, etc.) is associated with rapid economic growth and high demand for natural resources—witness China. However, as it is completed, the demand for natural resources diminishes. Assuming population levels stabilize, global demand could likewise stabilize. Capitalism supports rapid rates of technological change that improve productivity, hence providing more supply with fewer resources. One suggested solution to the ever-increasing amount of capital sloshing around the planet looking for a profitable investment is to use it to fund a renewable energy revolution (McCarthy, 2015).

2. A "fatal flaw" of technological fixes is that the solutions may induce their own environmental problems; e.g. the hydrofluorocarbons that replaced chlorofluorocarbons are also long-lived greenhouse gases. Technology creates complexity, and complexity may grow faster than the ability to understand and control it (Wylie, 1971).

EMT offers a systems-oriented approach to environmental science that aims to better integrate technology with the biosphere. It commits to a path of advanced technology solutions to ecological problems because they often open the only path to a sustainable future. An example is the recent international agreement that aims to phase out hydrofluorocarbons by means of improved refrigeration technologies.

3. If technological improvements increase energy efficiency, then energy prices will come down, but with the consequence that demand or consumption may well go up—an economic concept known as Jevons paradox (Alcott, 2005).

EM-based policies are supportive of government intervention in the market, as needed, to constrain aggregate increases in consumption (Wackernagel and Rees, 1997). This approach, of course, contravenes the proposition that the free market can solve all resource limitation problems. However, such interventions indeed take place when the environment is given full consideration in societal deliberations. The fact that Jevons paradox is widely applicable points to the need for systemic analyses (e.g., if the desired outcome is to reduce transportation-related carbon emissions, an analysis should consider whether the best policy is to legislate fuel efficiency or subsidize public transportation).

4. EMT was developed based on trends in several northern European countries and does not have general applicability. In the developing world, factors such as low availability of capital, limited effectiveness of governance, momentum of population growth, and low level of education all work against the EM project. Huber (2008, p. 365) suggests that "a country without sufficiently developed cultural and political coherence, institutional capacities, and especially... without state-society synergy on the basis of developmental politics rather than predatory ones, will not be able to successfully adopt complex new technology."

These are legitimate points, although perhaps most relevant to the worst-case situations. EM in Western Europe does depend on representative government, e.g., green political parties that advocate for pro-environment legislation. Globally, the number of democracies increased in the 1990s (mostly eastern European countries) and has been stable for about two decades. However, current trends are toward declines in political rights and civil liberties, which may weaken the possibilities for EM. The widespread adoption of cell phones in less developed countries is an interesting counterpoint to the view that less resource–demanding technologies cannot spread under non-western conditions.

5. The global economic system, or world system (Chirot and Hall, 1982), is configured in a way that prevents the spread of EM in the developing world. If indeed the global economy is structured in terms of a developed core that includes high-technology manufacturing, and a less developed periphery that supplies raw material and a market for manufactured products, then the countries in the less developed periphery are constrained from following the path of high-technology development that leads to EM.

The world system model of international development is now widely perceived as overgeneralized and superseded by many alternative paths to development; e.g., China, which is modernizing rapidly (Mol, 2006), at least economically, is not a core country.

6. Whatever success EM-inspired policies have had in reducing pollution locally comes by displacing it elsewhere (e.g., the increase in air pollution in China as economic globalization made it a manufacturing colossus).

The EM assumption is that as economic development succeeds (e.g., in China), national governmental and nongovernmental organizations will gradually reduce air pollution to improve quality of life. Recent policy changes in China seem to bear that out (e.g., a weakening commitment to coal and a strengthening commitment to renewable energy).

7. EM does not treat social justice issues associated with addressing global environmental change; it neglects power relations.

This is a challenge to the social sciences more generally (Castree, 2016; Hackmann, Moser, and St. Clair, 2014). EM as a theory of social change offers the opportunity for advancing (social) science through hypothesis formation and testing, e.g., by way of national-level case studies. The outcome of such research, if applied, may stimulate the EM process (Fisher and Freudenburg, 2001) or identify other routes to sustainable development.

Despite the many trenchant views of EMT, it has become a widely recognized paradigm for how humans might ameliorate the environmental costs of modernization and globalization (Mol, Sonnenfeld, and Spaargaren, 2009). Much of its appeal lies in offering a hopeful path. Critical perspectives on the role of capitalism and technology in driving global environmental change tend to be more concerned with documenting problems than with offering solutions. EMT provides a theoretical basis for the Great Transition. Case studies (albeit in the most developed world) demonstrating significant success are now available; e.g., the country of Denmark (Jamison and Baark, 1999) and the state of California in the United States (Schlosberg and Rinfret, 2008). Internationally, the number of nation-states with environmental ministries is increasing (Aklin and Urpelainen, 2014), which is the kind of institutional change EM would foster. The structure of the 2015 Paris Agreement on climate change is consistent with EMT in moving away from the top-down regulatory approach of the Kyoto Protocol, and casting climate change mitigation in terms of economic benefits (Dimitrov, 2016).

LONG-TERM EFFECTS OF ECOLOGICAL MODERNIZATION

The Great Transition (in the Anthropocene narrative) will require a restructuring of the relationship of the technosphere to the biosphere; i.e., integration rather than the current pattern of domination. Several changes, broadly consistent with EM, can be envisioned. Note that the prescriptions, as stated here, largely ignore significant normative challenges that would be associated with their implementation.

Finishing the Global Demographic Transition

Demographic studies of population growth rates at the national level reveal a broad trend of decreasing rates (Lutz and Samir, 2010). The most general explanation (albeit contested at times, Kulcsar, 2016) is found in demographic transition theory (Packard, 2016). The "demographic transition" refers to a change in population growth rate that typically accompanies economic development and modernization. Before development, rates of both births and deaths are high, and the population size fluctuates. In the early phases of development, improved health care reduces death rates, and the population growth rate increases. However, as general standards of living improve, both birth rates and death rates fall; thus, population size tends to stabilize. The underlying mechanism of falling birth rates is that as people become wealthier, they desire to have fewer children, and have the means to accomplish that goal. The desire for fewer children is based on declining concern about being taken care of in old age, increasing opportunities for women to have fulfilling careers, and recognition of the cost of raising and properly educating a child. The means to accomplish the goal of self-limiting fertility are the various technologies of birth control, notably the pill for women, which temporarily stops ovulation.

Government policies influence the demographic transition in widely varying ways, in some cases favoring it (e.g., social security and education of women) and in other ways impeding it (e.g., outlawing abortion). In the mid-twentieth century, development aid was often predicated on demographic transition theory, and hence included support for family planning. Beginning around 1980, however, neoliberal doctrine cast doubt on the assumption that decreasing the population growth rate benefits development (Kulcsar, 2016). Since then, support for family planning as a component of international development aid has dropped. Alternative views suggest that population policies can contribute to economic growth and limitation of carbon emissions in developing countries (Casey and Galor, 2017).

In any case, population growth rates in most countries in the developed world have passed through a demographic transition and, without accounting for immigration, would be stable or in decline. Japan, which limits immigration, now has a negative population growth rate (minus 0.2 percent per year). Germany, which allows significant immigration, has a low rate of growth (0.2 percent per year). The U.S. population would be stable without a high level of immigration.

In most developing countries, the demographic transition is well underway. However, in much of sub-Saharan Africa, birth rates remain relatively high. Ongoing economic development there is likely helping to slow birth rates, but formal governmental policies about population control could speed up the process. Unfortunately, political opposition (e.g., religious factions that oppose birth control on moral grounds) often impede governmental action. The politics of population control are also complicated by the possible perception of advocates as supporting inequity or as being racist.

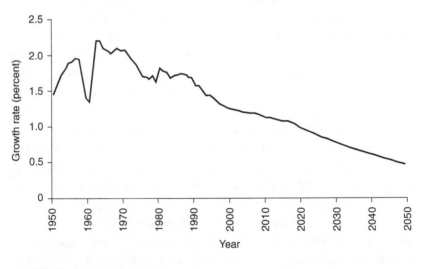

FIGURE 7.2

Historical and projected world population growth rates (USCB, 2017).

Globally, the Great Transition in the world population growth rates is clearly underway. The growth rate dropped from about 2 percent per year in the 1960s, to about 1 percent per year recently (figure 7.2). Mid-range projections suggest that the absolute number of people on the planet will top out at about 10 billion people around the turn of the century (figure 7.3). A peak at a lower figure is possible and would increase the likelihood of avoiding the BAU or Icarus scenarios (summarized in chapter 6). A key

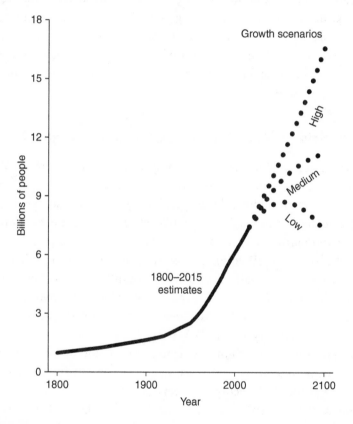

FIGURE 7.3

World population estimates from 1800 to 2100, based on "high," "medium," and "low" United Nations projections in 2015 and UN historical estimates for pre-1950 data. Graphic from "Projections of Population Growth," *Wikipedia*, https://en.wikipedia.org /wiki/Projections_of_population_growth. Data from the UN Department of Economic and Social Affairs, https://esa.un.org/unpd/wpp/.

observation is that 40 percent of pregnancies globally are not planned (Sedgh, Singh, and Hussain, 2014). Making pregnancy a conscious choice, and ensuring that women everywhere have the means to control their own fertility, would certainly speed up the global demographic transition. The path to that kind of world is broadly characterized as sustainable development (Sachs, 2015).

Although helpful in terms of reducing aggregate demand, and hence pressure on the environment, stabilization of population will be associated with significant social challenges, e.g., a declining ratio of working people to nonworking people. These are challenges to economists and social scientists. Immigration from areas of overpopulation could potentially ease demographic challenges in the developed world, if cultural barriers could be overcome.

Establishing a Global Network of Biodiversity Conservation Areas

We have noted the current high rate of species extinction (chapter 5), and the threat of much higher rates as the technosphere continues to expand and climate change alters the geographic range of each species (chapter 6). A globally coordinated effort (CBD, 2017) to create a network of terrestrial and marine conservation areas has begun and approximately 15 percent of land (UNEP, 2016), 13.5 percent of inland rivers (Abell, Lehner, Thieme, and Linke, 2016), and 4.5 percent of the oceans (Lubchenco and Grorud-Colvert, 2015) are so far in a protected status. Advocates argue that 25–50 percent of the land and ocean surface area will be required to minimize biodiversity loss (E. O. Wilson, 2016).

The political will to commit to habitat conservation is based on many factors. Economic benefits are cast in terms of increasing tourism, and provision of additional ecosystem services, such as water storage and filtration. Greater attention to "relational values" (i.e., values based on mutually beneficial relationships of individuals and societies to nature) could expand support for biodiversity conservation (Chan et al., 2016).

One link between conservation areas and ecological modernization is that support for conservation tends to increase as levels of income, education, and urbanization rise (Dietsch, Teel, and Manfredo, 2016). The proportion of the global population living is urban settings is increasing (now greater than 50 percent), and that proportion is expected to continue rising (Grimm et al., 2008; United Nations, 2005). The depopulated countryside and agricultural intensification offer opportunities for conversion of land use to conservation. However, to attain a net benefit in terms of conserving biodiversity, agricultural intensification must be done in an ecologically sound fashion, and in the context of landscape management (Matson and Vitousek, 2006). The land use footprint of cities is growing, but can be constrained to some degree by urban growth boundaries (Armstrong, Wang, and Tang, 2015).

Accomplishing a Renewable Energy Revolution

Revolution is the proper term here because the energy infrastructure is built into the fabric of society, and currently the energy infrastructure everywhere on the planet is solidly based on fossil fuels. The biosphere runs on solar energy; the technosphere runs on fossil fuel energy. Although increasing energy efficiency will help moderate fossil fuel emissions, the global demand for energy will likely increase for decades to come as economic development proceeds (Tainter, 2011), and as the waste products of the technosphere begin to be more comprehensively recycled (Haff, 2014). As we have seen in the BAU scenario, if the growing global energy demand is provided by the current reliance on combustion of fossil fuels, the stage is set for considerable human misery from climate change.

Carbon capture at the point of fossil fuel combustion (e.g., in power plants), combined with geological burial of the captured carbon, is possible. However, there are many impediments to operational facilities (IPCC, 2005) and investments in that direction are decreasing rather than increasing. Proposals have also been made to allow fossil fuel combustion, but to remove CO_2 directly from the atmosphere. Again, the technical constraints are significant and costs unknown (Broecker, 2008). To avoid the BAU scenario or worse, most of the energy required over the next century will need to be generated largely by hydroelectric, solar, wind, wave, geothermal, and nuclear sources.

The core requirement, of course, is an energy infrastructure that produces prodigious amounts of energy without emitting greenhouse gases or generating other undesirable externalities. It would take a lot of wind turbines (3,800,000), concentrated solar power plants (49,000), and solar photovoltaic power plants (40,000), along with ample hydroelectric facilities, geothermal plants, tidal turbines, and wave energy devices, but a global renewable energy revolution could be accomplished with mostly existing technology (Delluchi and Jacobson, 2011; Jacobson and Delluchi, 2009; Jacobson, Delucchi, Cameron, and Frew, 2015; Moriarty and Honnery, 2012). Issues of renewable energy intermittency and transmission are technically challenging, but potentially manageable. Ultra-high-voltage direct current transmission lines could link areas of high energy demand with remote areas of high renewable energy production. The transportation sector could rely on battery- and fuel cell–powered vehicles (e.g., supplied with hydrogen generated using renewable energy).

Economists are working on the kind of carbon taxes, feed-in tariffs, and energy technology research and development that will be required to accomplish a renewable energy revolution in this century (Bertram et al., 2015). Fruitful proposals about solving the international free rider problem (i.e., that some nations might refuse to reduce emissions while still benefiting from the mitigation actions of other nations) have also been developed (Nordhaus, 2015).

A first-order strategy with respect to electrical power generation would be to stop building fossil fuel–burning power plants. In the developed world, where energy demand is stable or decreasing, there would be a gradual turnover of existing power plants with replacement by renewable energy sources. In the developing world, only renewable energy plants would be built as energy demand grew.

Coal is the least efficient fossil fuel (i.e., energy delivery per unit of CO_2 emissions) and a significant step toward a renewable energy revolution would be to stop using it. Indeed, the construction of coal-burning power plants in the United States has stopped, although for a variety of reasons not all related to climate change. Unfortunately, the construction of new coal-burning power plants in the developing work is still common. This practice may decline rapidly as development agencies like the World Bank and the Organisation for Economic Cooperation and Development cease to fund coal-based projects (Bloomberg, 2015). The shift away from coal is also helped by the rapid decline in the cost of renewable energy, which is becoming competitive with coal in many regions. Global coal consumption is now projected to peak around 2020.

A switch from coal to natural gas has been proposed as part of the interim solution to rapidly reduce emissions. Natural gas delivers roughly twice the energy per unit of CO_2 emissions that coal delivers. And in the United States, it is relatively plentiful now because of advances in drilling techniques (i.e., hydraulic fracking). Natural gas power plants are also more compatible with renewable energy sources than coal-powered plants because they can be more readily switched on and off, hence helping with the intermittency problem. The benefits of natural gas are ambiguous, however, because even if global fossil fuel emissions could be cut in half it would not halt the rising atmospheric CO_2 concentration and risk of severe climate change impacts. There is also concern about contamination of local aquifers and leakage of methane from the natural gas mining and distribution infrastructure.

Remarkably, global carbon emissions were roughly stable in 2014, 2015, and 2016, something not previously seen outside a global recession (figure 7.4). The steady downward trend in carbon emissions per unit of economic activity is also encouraging. Much of a small 2015 emissions decline can be traced to the European Union, which has active policies to reduce CO_2 emissions (and was also suffering a mild economic downturn in that year) and to China, where coal consumption has apparently begun to decline (Jackson et al., 2016). It will be fascinating to monitor emissions in the coming years to see when a downward trend in global emissions actually begins. An inflection in the emission growth trajectory would certainly qualify as a marker for the Great Transition.

A major hurdle to the renewable energy revolution lies in generating the political will to make changes in public policy to support it. Ceasing to draw on the vast supply of readily available energy from fossil fuels will initially have significant economic costs. The policy community needs rigorous, consensus-based assessments of the feasibility of different paths to energy system transition. But these issues remain controversial (see, e.g., Jacobson et al., 2015, versus Clack et al., 2017). As to support from citizens, some commentators have evoked the analogy of a nation at war (McKibben, 2016). During wartime, citizens and government are focused

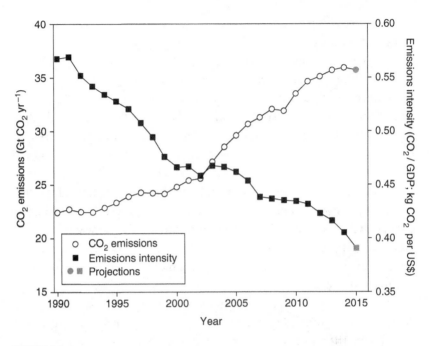

FIGURE 7.4

Global CO_2 emissions from fossil fuel use and industry since 1990 and emissions intensity CO_2/GDP (i.e., carbon emissions per unit of gross domestic product). The values for 2015 are late-in-the-year estimates. Adapted from Jackson et al. (2016).

on winning, and sacrifice is accepted. Increasing exposure to climate catastrophes may soon give the impression of a world at war and provide some needed bottom-up impetus.

World political leaders have been willing to invest some political capital toward reining in fossil fuel consumption and increasing renewable energy production. Virtually all countries that submitted plans (nationally determined contributions) to the recent Paris Agreement on climate change included provisions about increasing renewable energy. In the United States, at the initiative of the executive branch of the government, the Environmental Protection Agency sought to regulate CO_2 as a pollutant under the Clean Air Act. Unfortunately, changes in the political winds may impede that effort. Regulations to increase fuel efficiency in the transportation sector have also been promulgated widely. As noted,

China has officially begun moving away from reliance on coal. The Paris Agreement indicates a growing commitment across all nations to accomplish the renewable energy revolution.

High on the list of recalcitrant forces are the energy companies and nations that hold large reserves of coal, oil, and natural gas. The valuation of these companies (many publicly held) depends on the assumption that those reserves will be burned rather than "stranded" (i.e., left in the ground). There is not yet a model from political scientists and economists for dealing with this issue. Ecuador offered to forsake development of its oil fields in the western Amazon Basin if the country could be compensated. Unfortunately, there were no takers. Economists have noted that a moratorium now on new coal-fired power plants would help limit stranded assets in the future (Bertram et al., 2015).

From the perspective of the equilibration scenario (see chapter 6), a key requirement for human management of the global environment over the long term will be a capacity to alter the global climate. The climate may need to be warmed (e.g., to avoid an ice age), or cooled (e.g., if needed to offset a large geosphere source of greenhouse gases). We obviously know how to drive global warming by increasing greenhouse gas concentrations, but cooling the globe could require removal of greenhouse gases from the atmosphere. How that might be accomplished is now subject to intensive research (P. Smith et al., 2016), but almost any solution will likely require energy, and it must be renewable energy.

Greening of Governance at All Scales

Governance refers to collective efforts among individuals, civil society (e.g., NGOs), governmental agencies, and business interests to self-organize and self-regulate. Governance is green to the degree that issues of environmental quality come to influence decision-making processes. In chapter 8, we will examine the evolution of global environmental governance; in chapter 9, the kind of environmental monitoring needed to support global environmental governance; and in chapter 10, the development of socioecological systems at multiple scales. In a socioecological system, multiple stakeholders make consensus decisions about management of natural resources. The essential point here is that greening governance gives the environment a prominent place at the political decision-making table.

Greening of the Global Economy

Especially since the end of the Cold War in the late 1980s, capitalism in its many forms has dominated the global economy. Neoliberal economics, favoring open markets (free trade) and deregulation, has been ascendant. The threat of global environmental change now casts unbridled capitalism in a less favorable light. The rapid rise in greenhouse gas concentrations is essentially a massive market failure; the polluters (i.e., most of us) are not paying the environmental costs of greenhouse gas emissions. In many ways, we are seeing natural capital converted to economic capital, without regard to the fact that natural capital is nearly irreplaceable. As we have seen, EMT suggests that capitalism can be reformed. Green economy theory (Lamphere and Shefner, 2015) follows a similar line in countering neoliberalism by prioritizing activities of states, but it adds an element of social justice (e.g., creation of green jobs in the energy sector to help address inequality).

The process of greening global capitalism relies on the greening of governance, as previously described. A recent United Nations Environment Programme (UNEP) report on developing a green economy suggests it would be favored by (1) establishing proper regulatory frameworks; (2) prioritizing government investment and spending in areas that stimulate greening across economic sectors; (3) limiting government spending in areas that deplete natural capital; (4) using taxes and market-based instruments to promote green investment and innovation; (5) investing in capacity building, training, and education; and (6) strengthening international governance (UNEP, 2011). Corporations that operate globally could benefit from international agreements on environmental standards because it frees them from having to customize products for each regulatory regime.

Greening of the private sector would mean reform of both production and consumption. Greening the production supply chain refers to ensuring that the sources of raw material, transportation to the production facility, and the manufacturing process itself are environmentally friendly. For example, supply-chain governance in the soybean industry in Brazil has had significant success in delinking deforestation and soybean production (Gibbs et al., 2015). Life cycle analysis is beginning to trace the sustainability of all stages of production and consumption, including the final disposal of products (ideally recycled in one way or another rather than going to a landfill).

On the demand side, green consumers inform themselves about product options and buy the variants produced in the most sustainable fashion, albeit at perhaps a higher price than something produced as cheaply as possible. The question is not so much what is the absolute amount of consumption, but rather "what consumption is environmentally sustainable" (Mol and Spaargaren, 2004). Efforts by educational institutions, NGOs, and the media to inform consumers about availability of environmentally friendly products is crucial. In our enthusiasm for green labeling, however, we must not forget a parallel need to consider systemic change, e.g., by supporting improved public transportation as well as more efficient automobiles.

Changes toward greener capitalism will be driven by traditional regulatory approaches, but the corporate social responsibility movement (Carroll, 1999) is also providing significant impetus toward greener business operations (Lubin and Esty, 2010). For example, tech giant Google plans to run entirely on renewable energy by 2017. Especially in today's hyperactive media environment, a reputation for goodness promotes customer confidence and loyalty.

ESSENTIAL POINTS

The "end of history" has not truly arrived. The cultural evolution of western modernism and its spread by way of globalization has indeed produced a world that is increasingly integrated by market-based economics, albeit rather fractured politically. Modernization and globalization have also induced the Great Acceleration, which has brought the Earth system to a point at which the life support system for advanced technological civilization is threatened. Awareness of that threat has inspired a call for reflexive modernization, and a questioning of the assumptions of western modernism. EMT suggests the possibility of transitioning to a sustainable world. It relies on continued evolution of technology, on vibrant civil society engagement with environmental issues, on national-level participation in international environmental agreements, and on regulated capitalism. The latter will give national economies and the global economy the dynamism and vitality needed to reduce poverty (hence favoring the demographic transition) and to build a renewable energy infrastructure. EMT broadly predicts a greening of capitalism and a greening of governance. In many respects, Earth system scientists provide support for the EM process by informing societies of the actual condition and trends of the Earth system (especially as it is impacted by the technosphere), by indicating the risks of BAU scenarios, and by evaluating the possibilities for altering course.

IMPLICATIONS

1. Economic globalization often leads to conversion of natural capital into technosphere capital. This trend must be reversed to preserve the global life support system.

2. Globalization allows the uncoupling of economic activities and their environmental consequences (negative externalities). Corrective negative feedback loops are needed to make these linkages.

3. Modernization and globalization have tremendous momentum. That sociocultural and economic vitality must be harnessed to transform the global energy infrastructure into something renewable.

4. The global environment is being rapidly changed by the technosphere, and globally coordinated action on the environment is needed. Antiglobalization fosters a healthy reflexivity but tends to inhibit the nascent global community from acting on global environmental change issues.

5. Stabilization of the global population will help aid the transition to global sustainability, and thus governmental support for family planning remains desirable.

FURTHER READING

Chase-Dunn, C., and Lerro, B. (2013). *Social change: Globalization from the Stone Age to the present*. Boulder, CO: Paradigm.

Mol, A. P. J. (2000). Globalization and environment: Between apocalypse-blindness and ecological modernization. In Spaargaren G., Mol, A. P. J., and Buttel, F. H. (Eds.), *Environment and global modernity* (pp. 121–149). Thousand Oaks, CA: Sage.

8

GLOBAL ENVIRONMENTAL GOVERNANCE

*In the imagination of those who are sensitive to the realities of
our era, the earth has become a space ship, and this, perhaps, is
the most important single fact of our day.*

—K. Boulding (1965)

SUSTAINABLE MANAGEMENT OF NATURAL RESOURCES

Like other animals, humans consume natural resources. As commonly
defined, natural resources are substances or processes that humans did
not create but that can potentially benefit human welfare. Natural
resources are living, as with forests and grasslands, or nonliving, as with
rivers and streams. Natural resources derived from the biosphere are
inherently renewable, as is consistent with the flows of energy and cycling
of materials that characterize life. Natural resources from the geosphere,
such as coal and minerals, are often not renewable.

Our human ancestors generally extracted a small enough proportion
of their local natural resources that the resource base itself was not
degraded or depleted. However, as the global climate stabilized for the
Holocene interglacial period about 10,000 years ago, human hunting pressure began to reduce local herbivore populations. That impact on a local
resource put us on the path to cultivating plants and domesticating animals. The concentration of humans on resource-rich landscapes, and a

growth in technological capabilities, led to an increasing probability of natural resource degradation; i.e., the value of the ecological inheritance (Ellis, 2015) in those societies diminished over time.

As recounted in many excellent books (e.g., *The Earth as Transformed by Human Action* [B. L. Turner, Clark, Kates, and Richards, 1993]; *Collapse: How Societies Fail or Succeed* [Diamond, 2005]; and *Dirt: The Erosion of Civilization* [Montgomery, 2008]), recent human history is replete with cases in which technological advances in resource extraction ultimately degraded the local resource base. The flow of energy or materials that sustained the associated complex social structure was then reduced, and the social system collapsed. Alternatively, a society may not recognize natural resource limitations and thus overshoot the local carrying capacity. Historians continue to dispute the role of natural resource limitations in societal dynamics (e.g., Tainter, 1988), but clearly there are limits that are often poorly understood.

The alternative to resource degradation is sustainable management of natural resources. Sustainability means that resource capital is maintained even as a proportion of it is extracted and used for human purposes. Sustainability advocates emphasize both its economic and environmental aspects (Sustainable Development, 2017). The concept has applicability at scales ranging from the farm field, to the forest landscape, to the region, and to the planet. As we shall see, creating natural resource management schemes that achieve sustainability across all spatial scales is a tremendous challenge—certainly one to keep several generations of environmental and social scientists busy for decades or centuries to come.

The sustainability concept had its origin in the forest management literature relating to sustained yield of forest products. But it became especially important beginning in the 1990s in the context of economic development. In developing countries, environmental quality and natural capital were often being sacrificed to quickly generate financial and technosphere capital. To some degree, sustainability replaced the environmentalist's rationale of protecting the environment purely for its own sake, with the rationale of protecting the environment for the sake of human use (O'Neil, 2009).

What does sustainability mean at the global scale? A key question concerns the proportion of natural resources that can be redirected to support human purposes without compromising the functioning of the biosphere (McNeill, 2000). How much of the land surface must be left in a semi-natural state? As noted earlier, biologist E. O. Wilson recently (2016) proposed 50 percent. How much of global terrestrial net primary production can be appropriated (Foley, Monfreda, Ramankutty, and Zaks, 2007)? We are at nearly 40 percent now. How much of global fish production can be consumed? How much biodiversity can be lost?

These questions are just beginning to be addressed, often by discovering that some limit has been transgressed. Synthesis efforts such as the Millennium Assessment (2005) do an excellent job of summarizing the current state of affairs. Overall, we are doing abysmally (see chapter 4). To give only a few examples:

- Global forest area continues to decline. Only about a third of global wood production comes from planted forests. In tropical forests, what comes after logging of primary forests is commonly either a species-poor tree plantation or conversion to grazing or agriculture.
- Of the 42 percent of the global land mass that is classified as drylands, 60–70 percent of it is considered to have undergone some degree of desertification.
- Poor agricultural practices are associated with soil degradation and are converting large areas globally into unproductive wasteland.
- The rate of anthropogenically driven species extinctions is increasing rather than decreasing.
- The atmospheric carbon dioxide (CO_2) concentration is increasing at the rate of about 2 ppm (parts per million) per year, which presages rapid climate change. That change will exacerbate many natural resource management problems and deepen social problems associated with environmental refugees.

Progress made thus far toward global sustainability is more in the realm of conceptual development than in the trajectories of the various indices of Earth system condition. We are beginning to understand the scope of the problem and conceive of global-scale solutions, but implementing them lags far behind. The gap between what Earth system scientists perceive as the threats to humans from global environmental change, and the effectiveness of the existing environmental governance institutions at all scales, is growing.

PLANETARY BOUNDARIES

The recently developed concept of "planetary boundaries" (Rockström et al., 2009; Steffen et al., 2015b) helps to holistically frame the constellation of global environmental change issues. The boundaries here refer to specific properties of the Earth system that are undergoing rapid alteration by the technosphere. For each property, e.g., the atmospheric CO_2 concentration (linked to global mean temperature), a threshold value has been suggested beyond which the direction of change becomes practically irreversible, or a trajectory toward a new relatively stable state is locked in (table 8.1). The reference state is the Earth system in the 11,700-year-long Holocene epoch. This is the interval in Earth's history during which advanced technological civilization evolved. Crossing of the planetary boundaries will push the Earth system out of that comfort zone.

Given the uncertainties about how the global environment is changing, and our capacity to forecast the impacts of future changes, these thresholds have been characterized for now as more of a social construct than a strict limit (Biermann, 2012). Nevertheless, the concept has gained traction as a means to think about and discuss a "safe operating space for humanity."

PROGRESS TOWARD GLOBAL-SCALE ENVIRONMENTAL GOVERNANCE

Earth system governance has been defined as "the sum of the formal and informal rule systems and actor-networks at all levels of human society that are set up in order to influence the co-evolution of human and natural systems in a way that secures the sustainable development of human society" (Biermann, 2007, p. 329). In recent years, two models for global-scale environmental governance have emerged. The traditional model is a top-down approach in which nation-states negotiate international

TABLE 8.1 Planetary Boundaries

EARTH SYSTEM PROCESS	CONTROL VARIABLE(S)	PLANETARY BOUNDARY	VALUE IN 2009*
Climate change	Atmospheric CO_2 concentration	350 ppm	398 ppm
	Energy imbalance at top-of atmosphere	+1.0 W/m^2	+1.5 W/m^2
Stratospheric ozone	Stratospheric O_3 concentration, Dobson unit (DU)	< 5% reduction from preindustrial level of 290 DU	Only transgressed over Antarctica in Austral spring (~200 DU)
Ocean acidification	Carbonate ion concentration	≥ 80% of the preindustrial aragonite saturation state of mean surface ocean	~84% of the preindustrial aragonite saturation state
Biogeochemical flows (potassium [P] and nitrogen [N] cycles, in teragrams [Tg] per year)	P Global: Potassium flow from freshwater systems into the ocean	11 TgP/yr	~22 TgP/yr
	N Global: Industrial and intentional biological fixation of nitrogen	62 TgN/yr	150 TgN/yr
Land-system change	Percentage of global land cover converted to cropland	15%	12%
Freshwater use	Global: maximum amount of consumptive blue water use (km^3/yr)	4,000 km^3/yr	~2,600 km^3/yr
Atmospheric aerosol loading	Aerosol optical depth loading	Not specified	

(*continued*)

TABLE 8.1 Planetary Boundaries (*continued*)

EARTH SYSTEM PROCESS	CONTROL VARIABLE(S)	PLANETARY BOUNDARY	VALUE IN 2009*
Introduction of novel entities	Chemicals	Not specified	
Change in biosphere integrity	Extinction rate: Extinctions per million species years (E/MSY)	<10 E/MSY	100–1,000 E/MSY

*Values may have changed significantly since 2009.

Source: Rockström et al. (2009)

treaties and support regulatory bodies that manage resource use. In this model, international regulatory agreements are implemented mostly at the national level through executive authority or enforcement of legislated regulations. A second model is more informal and is situated primarily within the private sector; the governance arrangements involve mainly nongovernmental organizations and businesses. We will consider this second model first, and then survey what has been accomplished by the traditional top-down approach.

Bottom-Up Environmental Governance

Many forces currently contribute to an erosion of national-level sovereignty (Young, 1997). Economic globalization, internal strife along lines of religion or ethnic origin, the spread of human rights, and the necessity to address global change issues are leading examples. These trends drive the development of a more "bottom-up" approach to environmental governance. In contrast to the international and governmental flavor of the top-down model, this approach relies on an emerging network of transnational private sector organizations (Pattberg, 2007). This new form of natural resource management relies in part on business actors who are motivated by profit, but who understand that voluntary commitments to certain standards or practices will ultimately lead to sustained profits.

Often a nongovernmental organization (NGO) is the institutional actor that provides the coordinating links between multiple stakeholder groups associated with a particular resource.

That private governance arrangements can have large effects has been demonstrated strikingly in the forestry sector. The Forest Stewardship Council (FSC, 2017) links loggers, saw mills, merchants, consumers, and regulators. Because top-down regulation has been so ineffective in stopping deforestation, the FSC is a good case study of why the private governance approach is especially needed (Pattberg, 2007).

The economic leverage point at the heart of the FSC is the process of certification. Timber suppliers (public or private) that meet basic standards of sustainability (box 8.1), as established by an independent scientific organization, receive a periodic certification to that effect. Wood products are then labeled as certified, consumers seek out the suppliers of those products, and certification is thus rewarded. This self-regulatory process sidesteps the need for national-level regulation and enforcement. There are of course multiple issues regarding details of the certification

---

BOX 8.1

The Forest Stewardship Council Principles and Criteria for Sustainability

1. Compliance with laws
2. Workers' rights and employment conditions
3. Indigenous peoples' rights principle
4. Community relations principle
5. Benefits from the forest principle
6. Environmental values and impacts
7. Management planning
8. Monitoring and assessment
9. High conservation values
10. Implementation of management activities

(FSC, 2017)

process—notably defining what constitutes sustainable forestry—but this model for forest conservation has great potential.

Certification of fisheries and aquaculture facilities is also advancing rapidly. Criteria here (box 8.2) relate to conserving fish populations and ecosystems. The Marine Fisheries Council aims for a global scope and now certifies about 10 percent of total global wild-caught fish. Best Aquaculture Practices serves as a key certification body of cultured finfish, crustaceans, and mussels. As with forestry practices, there are complex issues with what constitutes sustainability in these contexts. Certification NGOs generally have formal objection procedures to maximize consensus, compliance, and transparency (C. Christian et al., 2013).

International certification of agricultural products has not proceeded in parallel with certification of forestry and fisheries products. National-level certification of organic agricultural products is done by agencies and NGOs in the United States, Canada, Japan, Switzerland, and counties in the European Union. They act largely independently of each other. Defining "organic" food products is even more fraught with issues than sustainable wood and fisheries products, notably with respect to tolerance for genetically modified organisms.

BOX 8.2

Marine Fisheries Council Principles of Sustainable Fishery

1. Maintenance and reestablishment of healthy populations of targeted species

2. Maintenance of the integrity of ecosystems

3. Development and maintenance of effective fisheries management systems, taking into account all relevant biological, technological, economic, social, environmental, and commercial aspects

4. Compliance with relevant local and national laws and standards, as well as international understandings and agreements

(MFC, 2016)

The certification movement has also been successful in the construction industry, with the Green Building Council (USGBC, 2017) as the most active building certification NGO in the United States and internationally. USGBC has established the Leadership in Energy and Environmental Design (LEED) label as its stamp of approval. Criteria for certification relate to site characteristics, energy efficiency, source and type of building materials, and design features. Building certification generally contributes to the value of a building and is commonly required for publicly funded buildings.

The diversity and influence of environmental NGOs has increased tremendously in recent decades. This influence is facilitated by the Internet, which enhances information flow within and across national borders. NGOs lobby, educate, manage natural resources, and take direct action against polluters. Collectively, the array of environmental NGOs contributes to the emergence of a global civil society (Lipschutz, 1996).

Top-Down Environmental Governance

The nation-state is the fulcrum point for internationally coordinated efforts toward global environmental governance. States send representatives to international negotiating bodies to work out mutually agreeable strategies, and within their own borders implement supporting policies. States have traditionally been most concerned with defense, economic growth, and social welfare. However, environmental quality (both local and global) has emerged as a critical new responsibility (Eckersley, 2004).

WORLD TRADE ORGANIZATION, INTERNATIONAL MONETARY FUND, AND WORLD BANK

In addition to the United Nations–mediated activities discussed later in this chapter, governmental development agencies, international trade agreements, and foreign aid programs exert a significant influence on global environmental quality (Clapp and Dauvergne, 2005). The World Trade Organization (WTO) is an international institution (153 countries) that aims to develop trade agreements. It is a successor to an influential post–World War II global trade agreement (the General Agreement on Tariffs and Trade [GATT]). The WTO originated in the period of 1986–1994 and provided the institutional home for negotiations that led to regional agreements such as the North American Free Trade Agreement (NAFTA). The WTO has become an arbiter of environmental quality on various issues because it must decide when tariffs can be established based on environmental standards. One benefit of WTO emphasis on trade expansion is that exporting countries may adopt higher environmental standards to compete in markets where those standards are the norm. It is now common, when negotiating international trade agreements, to address issues related to impacts of trade on environmental quality.

The overall costs and benefits of WTO agreements for the environment remain controversial. Binding provisions to protect the environment are often included, but enforcement has been limited. WTO agreements can work against environmental protection in cases where corporations sue governments based on provisions that prevent regulation of investments based on environmental issues. The latest round of WTO negotiations (the Doha round) has been especially contentious. This pattern is related in part to push-back from emerging economy countries, notably China, India, and Brazil, about free trade provisions that would open the U.S. agricultural market (Hopewell, 2016). Anti-globalization sentiment in the most developed countries (see chapter 7) has also contributed. Failure to agree on a global trade rule set with significant support for beneficial environmental practices is a missed opportunity with respect to addressing multiple aspects of global environmental change.

The International Monetary Fund (IMF) and World Bank are additional international institutions with noteworthy influence of environmental quality. These organizations were organized in 1944 to provide advice and financing to their member countries. The initial focus was on rebuilding after World War II. However, by the 1990s, these organizations had begun to consider environmental issues and to promote sustainable growth. Structural adjustment lending (SAL) is a key means by which they influence environmental quality. These nation-level lending agreements are contingent upon policy reforms, which often have environmental implications (direct and indirect). The World Bank may fund development of a National Environmental Action Plan as part of negotiations toward approval of projects with significant environmental impacts. On the negative side, mandated reductions in social spending associated with SALs tend to increase poverty, thus pushing the poor to exploit available natural resources. Decreased government spending may also reduce the effectiveness of the local environmental regulatory agencies. SALs encourage development of export crops, which may be problematic environmentally (e.g., the relationship of soybean export to deforestation in Brazil).

In response to NGO pressure, the World Bank has moved to "green up" its portfolio of international projects, e.g., by providing less lending for megadams (World Bank, 2013b). The World Bank has also been a long-term supporter of population control projects (albeit with considerable controversial at times), thus fostering the global demographic transition.

UNITED NATIONS

The history of United Nations (UN)–mediated global environmental governance of bodies and agreements outside the IMF and World Bank framework is rather short, so it is worth going back to the beginning. This history could logically start with the League of Nations, which was born out of the debacle of World War I. However, the League of Nations apparently did not have an environmental component, and in any case died prematurely because of its failure to deal substantively with the geopolitical aftermath of the war. Next up was the UN. In its initial form after WWII, there was little of substance related to environmental issues. But over the years it has come to play a major role in international negotiations on the environment.

The UN-sponsored Stockholm Conference in 1972 marked the beginning of broad international recognition of the significance of global environmental change issues. The primary outcome of the conference was the nonbinding Stockholm Declaration. Fundamentally, it established the principle that each country was responsible for the human-induced environmental changes within its borders that affected the environment outside its borders. The conference also initiated the formation of the UN Environment Programme (UNEP), thus creating a forum for ongoing discussions of global environmental change issues.

Twenty years later (1992), the UN-sponsored the Conference on the Environment and Development (UNCED) in Rio de Janeiro. Popularly known as the "Earth Summit," this was a massive conference involving

100 heads of state, representatives of 178 countries, and representatives of 1,000 NGOs. Some of the agreements reached in this meeting resulted in binding commitments, notably the Framework Convention on Climate Change (discussed below). It required countries to regularly produce estimates of greenhouse gas emissions. Other important outcomes were a Framework Convention on Biodiversity, a Convention on Desertification, and a Statement of Principles on Forests.

UNCED initiated massive participation by NGOs in the development of international environmental agreements. The NGOs included advocacy groups that worked primarily in the political arena, as well as educational groups, environmental activist groups like Greenpeace, and certification proponents like the Forest Stewardship Council.

The follow-up international meeting to UNCED was the UN World Summit on Sustainable Development held in Johannesburg, South Africa, in 2002. It is widely considered a failure. Agendas for change had already been set in previous environmental megaconferences, and it was time for action plans. Unfortunately, the necessary political will could not be mustered (Seyfang, 2003). Likewise, the Rio 2012 conference held in Rio de Janeiro in 2012 generated a variety of nonbinding resolutions that have not had much impact. There is growing skepticism about the effectiveness of the UN consensus process.

The UN framework for global environmental governance has had the effect of promoting, in many countries, environmental quality regulatory bodies or environmental ministries (Aklin and Urpelainen, 2014) that support international negotiations about environmental issues (i.e., multilateral environmental agreements). The U.S. Environmental Protection Agency (EPA) is perhaps the prototype. It was established in 1970 to protect the environmental and human health. EPA now works extensively on global environmental change issues, including stratospheric ozone depletion and climate change.

Domain-Specific International Agreements

Agreements Related to the Oceans

Oceans provided the impetus for one of the first international environmental agreements. In 1937, an agreement was signed in London that began the effort to regulate whale harvesting. A successor convention in 1946 established the International Whaling Commission (IWC). The governing protocol was a simple majority rule, and since commercial whaling was widely accepted and enforcement of commission regulations was weak, decline in whale populations continued. By 1992, a block of anti-whaling countries managed to collaborate and pass a moratorium on commercial whaling. Provisions included exemptions for purposes of scientific research and the option to gain an exemption by registering an objection. Thus, a small group of countries continued to hunt whales. The IWC is now at something of an impasse between the pro-whaling and anti-whaling blocks of nations. Nevertheless, whale hunting has declined dramatically since the 1960s—driven in part by direct action from environmental NGOs against whaling ships—and populations of many whale species are increasing.

Besides whaling, the issue of dumping at sea has stirred action. An international convention in 1972 prohibited dumping material that was potentially hazardous to marine life. (Dumping of anything else was acceptable.) The convention was updated in 1996 to prohibit dumping of any waste. Most recently, the UN has developed a Convention on the Law of the Sea that encompasses a wide variety of environmental issues. Enough countries signed the convention to bring it into force by 1994 for all parties to the treaty. The United States has signed onto the Law of the Sea Convention, but it has not been ratified by the U.S. Senate, mostly because mining interests are resisting regulation of deep sea mining. Other questionable features include a provision that if a nation does not harvest its entire allowable catch of fish, it must give rights to the surplus to other nations.

Agreements Related to the Atmosphere

Radionucleotides

The first international treaties regarding the atmosphere concerned the issue of radioactive contamination. As nuclear testing escalated in the 1950s, it was already known that detonation of nuclear bombs above ground produced radioactive substances in the atmosphere that were potentially hazardous to human health. A nuclear test explosion at Bikini Atoll in 1954 was two or three times as powerful as scientists had predicted, and the radioactive products of the explosion reached a nearby inhabited atoll, causing radiation sickness in the human residents.

By the late 1950s, the United States and the Soviet Union were exploding nuclear weapons in space to test their potential effectiveness as anti-satellite weapons. Incredibly, both nations set off nuclear weapons in the atmosphere as a form of saber rattling during the Cuban Missile Crisis in 1963. That incident must have generated a keen sense in the international community that something more had to be done to step back from the nuclear brink. One step in that direction was the Test Ban Treaty in 1963, ostensibly driven by the increasing concentration of radioactive substances in the atmosphere. This agreement prohibited nuclear tests in the atmosphere, space, and underwater; the goal being "an end to the contamination of man's environment by radioactive substances." In 1996, five of the eight nations with nuclear weapons capability signed a Comprehensive Test Ban Treaty. India, Pakistan, and North Korea did not sign but have limited themselves to underground explosions, again often as a form of saber rattling.

Sulfur and Nitrogen Emissions

Sulfate and nitrate emissions from fossil fuel combustion are another form of air pollution that has inspired international treaties (beginning in the 1970s). These compounds eventually come out of the atmosphere by way of precipitation and contribute to soil and water acidification (as outlined in chapter 4). In the early 1970s, it became evident that emissions from England that were blown east by the prevailing winds were acidifying lakes and streams in Scandinavia. Likewise, emissions from industrial facilities in the U.S. Midwest were driving lake acidification in Canada. Negotiations on these issues resulted in the Convention on Long-Range Transboundary Air Pollution, signed in 1979. The original convention did not quantify how much air pollution a given country could emit, but follow-up agreements have become increasingly specific. The transboundary emissions were mostly associated with power plants, and technologies for reducing emissions included scrubbers that stripped out the sulfur.

On a personal note, I did a postdoctoral fellowship at an EPA laboratory in the mid-1980s studying the effects of acidic fog on leaching of nutrients from conifer foliage. Near the end of my study, the United States passed legislation to limit emissions of sulfur from U.S. power plants and subsequently, these emissions began to decrease. As soon as the legislation was passed, the EPA largely lost interest in acid rain, and we researchers joked about the problem being "solved." In reality, the level of sulfur deposition has dropped in the United States, but levels of nitrate deposition have remained high in many areas. Acidic deposition in China and India is clearly degrading environmental quality. Throughout the biosphere, serious questions remain about the long-term impacts of air pollution on acidification of surface waters and soil.

Stratospheric Ozone Depletion

The role of chlorofluorocarbons (CFCs) in reducing ozone levels in the stratosphere is another critical atmospheric pollution problem (see chapter 4) that has generated an international treaty. CFCs were first synthesized in the 1930s for use as refrigerants and propellants. The science community was alerted to a possible problem when James Lovelock (of Gaia fame) pointed out in 1972 that CFCs were accumulating in the atmosphere. He was casting around for a scientific question that could be addressed by a new instrument he had designed called an electron capture device. It was capable of measuring concentrations of trace gases such as CFCs at unprecedentedly low levels. He used it on a scientific expedition to Antarctica in 1972 to begin monitoring CFC concentrations. After Lovelock's discovery, atmospheric chemists James Molina and Sherry Rowland hypothesized that accumulating CFCs might interfere with ozone chemistry in the atmosphere. By 1985 measurements of ozone in the stratosphere over Antarctica had confirmed declines during the winter. The mechanisms related to an increase in ice crystals during the winter that helped catalyze ozone-consuming reactions. Less stratospheric ozone meant more solar ultraviolet (UV) radiation penetrated the atmosphere. UNEP organized a scientific working ground on the issue, and in 1985 an international accord was signed. It did not contain specific constraints on CFC production, but included a commitment to "protect humans from harm associated with human-induced changes in the ozone layer." A scientific consensus rapidly built up about the issue, and the chemical industry developed alternative compounds that could substitute for

CFCs. The Montreal Protocol in 1987 introduced firm reductions in CFC production.

I was still working at an EPA laboratory about this time (late 1980s) and became involved in assessing possible ecosystem-level effects of increasing UV radiation. Scientists were performing experiments such as growing tree seedlings under lamps with special radiation regimes mimicking the case of severe stratospheric ozone depletion. From these studies, it was evident that certain plant pigments help protect the plants from UV radiation, and that concentrations of the pigments rose as UV exposure rose. Plants growing at high elevations, with a thinner atmosphere and hence relatively high levels of UV radiation, tended to have higher concentrations of the protecting pigments. In some studies, plant productivity declined under exposure to elevated levels of UV radiation (specifically in the UV-B wavelengths). While these studies were progressing, the international policy community had rapidly come together. The Montreal Protocol, which went into force in 1989, largely stopped further production of the specific CFCs that were causing the ozone depletion. As far as further research, it was again a case of "problem solved." The EPA got out of UV effects research. Only recently are we beginning to see recovery of stratospheric ozone (P. J. Nair et al., 2015) as concentrations of chlorofluorocarbons drop (see figure 4.5)

The Montreal Protocol is often highlighted as a model for addressing other global environmental change issues, notably climate change. Especially regarding reliance on interdisciplinary science to assess the issue, and the governance approach to addressing the stratospheric ozone depletion issues, it did indeed set useful precedents. However, the fact that relatively few CFC producers were involved, and that CFC substitutes were rapidly developed, helped constrain the problem. Responding to climate change presents a qualitatively different case because of the pervasive use of fossil fuels.

Interestingly, the Montreal Protocol is helping with the climate change issue because it has become the vehicle for efforts to reduce emissions and concentrations of hydrochlorofluorocarbons (HCFCs). These compounds were developed to replace CFCs when stratospheric ozone depletion became an issue. Unfortunately, they are also strong greenhouse gases. Another generation of alternative compounds has been developed, and most countries are on board to add provisions to the Montreal Protocol that formally ban HCFC emissions.

Agreements Related to the Land

Issues related to the land tend to be more localized than those related to the oceans or atmosphere. Consequently, it has taken somewhat longer for them to be addressed at the international level. The land issues that have now come to the forefront are deforestation, desertification, and loss of biodiversity. All three were addressed with formal conventions or declarations at UNCED.

The UNCED Convention on Biodiversity (UCB) established biodiversity as a core environmental issue. Unfortunately, the UCB lacked mechanisms for enforcement of its guiding principles and has therefore had limited effectiveness. It raised many thorny issues within the biodiversity sphere—notably the intellectual property rights related to biological and genetic resources, and the role of indigenous people in maintaining biodiversity. Nonetheless, the UCB has at least provided a framework for ongoing conservation efforts. The UCB declared a formal goal of reducing the global loss of biodiversity by 2010 (not achieved). An additional formal goal was to protect at least 10 percent of land within each biome or ecoregion by 2010, which likewise has not been achieved (Jenkins and Joppa, 2009).

The UNCED Declaration of Principles on Forests is a rather weak incentive to forest conservation relative to the massive ongoing assault on tropical forests. Follow-up efforts, such as the Tropical Forestry Action Plan and the United Nations Forum on Forests, have also had little practical effect. The worst deforestation occurs in tropical countries, where development has a much higher priority than conservation.

One hopeful sign is the linkage of deforestation with the climate change issue. Current proposals call for payment (carbon credits) to countries with areas of intact tropical forest to keep them from being deforested (the acronym is REDD: Reduction in Deforestation and Forest Degradation). REDD+ is an elaboration of REDD allowing activities for enhancement of forest carbon stocks as well as prevention of carbon losses by deforestation.

The UN Convention to Combat Desertification (UNCCD) was also fundamentally a statement of principles rather than a set of enforceable regulations. It was primarily designed to bring the issue onto the global stage. As with deforestation, problems resulting from development are at the forefront of proposals to reduce desertification. UNCCD has been a force for development of national-level action plans and for promotion of research that identifies the underlying causes of desertification. These factors include notably complex issues such as insecure land tenure, failed local institutions, and low market prices.

To serve as a financial mechanism linking international agreements on the environment with on-the-ground projects, the Global Environmental Facility (GEF) was created in 1994 under the auspices of UNEP, the World Bank, and the United Nations Development Programme. Donor countries have contributed billions of dollars to support projects in developing countries related to the UCB, UNCCD, and other conventions. GEF is active in transboundary natural resource management (e.g., the Mekong River Basin) and global-scale conservation efforts such as the World Database of Protected Areas developed by the International Union for the Conservation of Nature.

ADDRESSING GLOBAL CLIMATE CHANGE

By the late 1980s, growing awareness of rising concentrations of CO_2, methane, nitrous oxide, tropospheric ozone, and CFCs had raised the possibility that associated global climate warming might profoundly impact human welfare. The international response was the founding of the Intergovernmental Panel on Climate Change (IPCC). This body was established in 1988 under the auspices of the World Meteorological Organization and UNEP. The IPCC aimed to produce a consensus among scientists across an array of disciplines about the issue of climate change.

The first of a series of IPCC assessment reports was released in 1990 and the most recent in 2014. The 1990 document concluded that climate change, driven by increasing concentrations of greenhouse gases, was indeed a threat. In a truly impressive testament to the emerging global awareness of environmental issues, the UN General Assembly initiated international negotiations on the climate change issue based on the IPCC report. By 1992, the Framework Convention on Climate Change (FCCC) was opened for signatures.

One of the most important provisions of the UNFCCC was that each country develops an annual inventory of its greenhouse gas emissions. These emission values established the basis for the follow-up Kyoto Protocol, which specified country-level targets for reduction of greenhouse gas emissions. Enough countries (albeit not including the United States, which was the largest single greenhouse gas–emitting country at the time) eventually ratified the Kyoto Protocol, and it went into force in 2005. The first target dates for evaluating the commitments to emissions reductions were in 2012. By that time, large emitters (like Canada) that had signed on but were not achieving their emissions reduction goals withdrew to avoid financial sanctions. The Kyoto Protocol succeeded as a prototype attempt at an international agreement but was a failure in terms of actually reducing global greenhouse gas emissions.

There were initially high hopes for a successor treaty to be formulated at the climate summit in Copenhagen in 2009. However, insufficient political will was available to get much done. After the disappointment at Copenhagen, some policy analysts thought it might never be possible to reach agreement among all nations about how to address global climate change. An alternative might be an agreement among the ten or so largest emitters of greenhouse gases, notably China (which has surpassed the United States in greenhouse gas emissions), India (also rapidly developing, and with an ample supply of coal), and the United States. Those three countries contribute a substantial proportion (nearly 50 percent) of the total global emissions (Boden, Marland, and Andres, 2015) and might be more likely to act in a coordinated fashion (Levi, 2009). Economist William Nordhaus has proposed that a "climate club" of a few high-emitting countries could agree to limit emissions and impose small trade penalties to induce other countries to participate (Nordhaus, 2015).

One thing on which the Copenhagen meeting participants did agree was the desirability of reducing deforestation (a significant source of CO_2). The REDD+ concept was supported by commitments of the developed countries to help pay for relevant programs. Norway committed funds equivalent to half a billion U.S. dollars per year to the program. The straightforward objective of reducing emissions through REDD+ is complicated by issues such as rights of indigenous people, corruption in the granting of logging concessions, "leakage" in the sense of conservation in one area contributing to exploitation in another, and development of systems for monitoring carbon stocks and flux.

The most recent phase of the international negotiations on CO_2 emissions, the Paris Agreement, was brought to fruition in December 2015. Several key differences from previous negotiated agreements were manifest.

1. All 195 countries participating in the negotiations approved the final working of the agreement.

2. A target of less than a 2.0°C increase (by 2100) in global mean temperature above the preindustrial level was specified.

3. Each county specified its own "nationally determined contributions" to emissions reductions rather than having its commitments determined by negotiation. This was a change away from the top-down model of previous attempts. Correspondingly, there was no agreed upon enforcement mechanism, simply a plan to assess overall progress toward emissions reductions every 5 years.

4. The importance of forests as carbon sources and sinks was given special emphasis.

5. The subject of adaptation was raised, a recognition that significant climate changes will occur no matter what the level of mitigation effort.

The primary criticisms of the Paris Agreement center on its nonbinding nature, its modest target of 1.5°C–2.0°C (Schleussner et al., 2016), and the gap between the sum of the national-level commitments, and what would realistically be required to meet the less than 2°C target. Note that for the first half of 2016, global mean temperature was 1.9°C above the twentieth century average value (NOAA, 2017a).

Despite the nonbinding nature of the Paris Agreement, in early 2017 the president of the United States chose to withdraw U.S. participation. This symbolic gesture from the nation with the largest contribution to cumulative fossil fuel emissions inspired near-universal opprobrium. Ironically, for a president concerned with trade imbalance, U.S. withdrawal could precipitate a carbon tax on U.S. imports in other nations (Kemp, 2017).

Besides the international negotiations, efforts to reduce CO_2 emissions are proceeding in many cases at the national level and lower, notably in legislation that supports development of renewable energy sources and research on carbon capture (from coal-burning power plants). There has been less support for carbon taxes, although significantly Australia passed one (November 2011, later canceled). California passed a remarkably progressive climate change bill in 2006 that implements a "cap-and-trade" system for reducing greenhouse gas emissions (and indeed total emissions are going down). Bottom-up pressure from NGOs such as 350.org has lent impetus to reform efforts.

We noted in chapter 4 that the atmospheric concentrations of methane and nitrous oxide were increasing along with CO_2. The possibility of methane emissions, associated with the transition away from coal toward natural gas, is particularly worrisome. The global policy community has not made much progress on an international agreement to limit methane emissions, and national-level efforts in the United States to upgrade standards on methane leakage from natural gas facilities have been weakened. The possibilities for meeting the goal of keeping the global mean temperature increase below $1.5°C–2.0°C$ will be significantly improved if the concentrations of methane are controlled along with CO_2 (Shindell et al., 2012)

The climate change issue raises many fundamental questions about how humanity might organize itself to address global environmental change problems (Glover, 2006). There are obvious equity concerns given that developed countries have contributed much more to the current increase in greenhouse gases than developing countries, whereas the developing countries will need substantial new energy sources to improve their economies and standard of living, and have less capacity to adapt. The quasi-advisory role of science in societal deliberations about climate change is increasingly different ("post-normal," box 8.3) than its traditional role of making hypotheses and testing them with observations ("positivist science"). Nation-states have decision-making power regarding climate change mitigation (e.g., the UNFCCC and the Paris Agreement), but do they represent all interests? This bewildering array of new

BOX 8.3
Post-Normal Science

The concept of "wicked problems" arose in the 1970s (Rittel and Webber, 1973), and the related concept of "post-normal science" in the 1990s (Funtowicz and Ravetz, 1994). Wicked problems commonly have the following features (Ney and Verweij, 2015, p. 1679):

> (1) each wicked problem is, in important respects, unique; (2) the range of possible causes is large and uncertain; (3) the set of solutions is equally vast and open-ended; (4) many people and organizations, from various social and natural domains, are involved; (5) each solution requires investing lots of time, energy and money resulting in large-scale changes in behavior, ecosystems, infrastructure and technology; (6) implementing any solution likely creates novel problems; and (7) as wicked problems are multi-faceted and enduring, it is inappropriate to speak of correct solutions in any absolute sense—it is preferable to use relative terms, such as satisfactory, more helpful or more widely acceptable.

Traditional pure and applied sciences have difficulty addressing wicked problems in part because values are often involved in their solution, something inconsistent with the putative objective nature of science. The "extended peer review community" for wicked problems includes scientists from very different disciplines, as well as nonscientist with direct experience of the problem (Funtowicz and Ravetz, 1994). Thus, achieving consensus in the usual scientific way becomes more difficult. The emerging social contract of science and society calls for scientific help in addressing wicked problems (Lubchenco, 1998); however, there has been a corresponding increase in awareness of the limits of scientific prediction and control. Scientists are participants, but no longer privileged authorities, in the intensive dialogues needed to navigate toward solutions to wicked environmental change problems.

perspectives indicates the still open and contested nature of the search for solutions to global environmental change problems. Even the concept of the global "we," who have ushered in the Anthropocene and now must begin managing the Earth system, is contested (Malm and Hornborg, 2014).

REQUIREMENTS FOR EFFECTIVE GLOBAL ENVIRONMENTAL GOVERNANCE

Considering the whole array of global environmental change issues, a few generalized requirements for progress in global environmental governance have been recognized.

1. *The capacity to monitor the resource* (see chapter 9). Status and trends of the resource provide a basis for negotiating levels of intervention and restrictions. This capacity at the global scale is technically challenging and not well coordinated. It requires national and international research institutions focused on monitoring.

2. *The capacity to forecast the resource condition based on simulation modeling* (see chapter 6). The projection of current trends into future states with and without policy changes supports negotiations on global environmental change issues.

3. *Quantitative targets for indices of environmental quality* (e.g., planetary boundaries). A target value helps bring into focus what kinds of changes are needed to avoid dangerous, virtually irreversible, changes.

4. *Enforcement of negotiated agreements.* Often global environmental change agreements are aspirational rather than enforceable. National sovereignty generally trumps global environmental quality. Economic sanctions are currently used among nations in attempts to force compliance with global standards on issues such as human rights and security. Ultimately, this approach might prove successful with respect to compliance on global environmental change issues.

5. *A forum for stakeholder engagement.* UNEP now provides a first-order forum for global environmental change issues. However, the complexity of the issues means that multiple levels of governance (global, national, state, etc.) must be engaged and multiple organizational

objectives must be coordinated (Galaz, Biermann, Folke, Nilsson, and Olsson, 2012).

6. *An informed population* that understand the issues and has access to enough information and assessments to help provide the political will to make appropriate policy changes and participate in NGOs that advocate for reform (Murtugudde, 2010).

The massive scale of global environmental change issues and the slow progress toward addressing them has inspired the political science community to debate the optimal form of global environmental governance institutions. Some have pushed for an upgrading of UNEP to the level of a WTO-like organization within the UN framework, or to creation of a new coordinating institution. Biermann (2012, 2014) has long advocated for a World Environmental Organization (WEO). This intergovernmental institution would be designed along the lines of the WTO, which has become one of the major institutions of global economic governance. The benefits of an overarching global environmental change institution include possible synergies in linking governance across the range of planetary boundaries (Nilsson and Persson, 2012). Others have questioned the possible effectiveness of a WEO, especially considering the tight grip on sovereignty demanded by most nations (Biermann, 2014).

ESSENTIAL POINTS

Traditional top-down environmental governance organizations (largely based on the United Nations) have succeeded in formulating a wide array of multilateral environmental agreements aimed at sustainable management of global natural resources. However, implementation of these agreements has been patchy, often running up against intractable issues of equity. NGOs such as the Global Environmental Fund help coordinate bottom-up actions to support the goals of the international agreements.

Bottom-up efforts at global environmental governance—notably NGOs that certify forestry, agriculture, and fisheries operations—are creating an institutional infrastructure that links producers, consumers, and regulators. The higher the proportion of global natural resources that are certified, the better the chances for reversing negative trends in environmental quality. Environmental NGOs also lobby, educate, manage natural resources, and take direct action against polluters.

Although the 2015 Paris Agreement established a global consensus that greenhouse gas emissions must be reduced, the summed national-level commitments are likely not sufficient to accomplish the stated goal of keeping mean temperature increase below 1.5°C–2.0°C. A change of 2°C or greater will have massive impacts on the global biogeochemical cycles, the biosphere, and human welfare.

IMPLICATIONS

Although WTO-sponsored negotiations have stalled in recent years, issues relating to trade and the environment are becoming more closely linked. Global, regional, and bilateral trade agreements are an opportunity to standardize progressive environmental practices.

A new global environmental governance institution (i.e., a WEO) could strengthen the relevant scientific knowledge base and facilitate international cooperation on global environmental change issues. Conservation, mitigation, and adaptation could all fall under its purview.

It will be important, in follow-up meetings to the Paris Agreement on climate change, to make additional commitments to reduce net greenhouse gas emissions.

FURTHER READING

Biermann, F. (2014). *Earth system governance: World politics in the Anthropocene.* Cambridge, MA: MIT Press.

Young, O. R. (Ed.). (1997). *Global governance.* Cambridge, MA: MIT Press.

9

GLOBAL MONITORING

Understanding changes in climate, biodiversity, and our knowl-edge of and ability to predict natural hazards requires long-term observations, and the return on investment for these systems to local, state, and federal governments and to the public at large is immeasurable.

—Scripps Oceanography (2015)

INSTITUTIONAL ASPECTS

A recurrent theme in the international environmental protection agreements discussed previously is the need for monitoring. It serves multiple functions. First is to raise awareness of new issues. Scientists originally became aware of the increasing concentration of carbon dioxide (CO_2), and the depletion of stratospheric ozone, thanks to monitoring programs that had been maintained long enough to detect a trend or change. Second is providing the scientific basis to inspire changes in natural resource management. For example, dire declines in the world's terrestrial megafauna, as documented by periodic updates of the International Union for the Conservation of Nature (IUCN) Red List of Endangered Species (IUCN, 2017), are highlighted in calls for policy changes to preserve biodiversity (Ripple et al., 2016). A third function is confirmation of compliance with national laws or international agreements. In Brazil, it is only recently that satellite-based near real-time monitoring of deforestation has become operational, thus offering the possibility to effectively enforce national laws protecting the forest (Tollefson, 2015).

Monitoring not only tells us if an environmental problem is getting worse, but also if it is getting better. Repeated surveys of the grey whale population on the West Coast of the United States have revealed a steadily growing population associated with the implementation of conservation efforts over the whole north–south range of the species. The network of precipitation-monitoring sites maintained by the U.S. National Atmospheric Deposition Program captured the decline in deposition of sulfur and nitrogen in the eastern United States in response to the Clean Air Act and its amendments. Monitoring of stratospheric ozone now suggests the beginning of a recovery (Solomon et al., 2016).

The United Nations has been a strong advocate for global monitoring, but operational programs are still quite limited. The UN Environment Programme (UNEP), established in 1972, has a mandate to monitor the global environment, but its performance has been problematical (Ivanova 2005). The UN Food and Agriculture Organization (FAO) has historically monitored global crop, forestry, and fishery production by assembling national-level data into global summaries. FAO has more recently supported the formation of an Integrated Global Observing System (IGOS), which aims at comprehensive monitoring of the climate, oceans, and land. IGOS is broken out into about 20 subsidiary organizations, e.g., the Global Climate Observing System (GCOS, 2017a).

These new organizations generally do not collect and analyze data themselves. Rather they aim to stimulate and coordinate national-level efforts. The goal is to provide relevant environmental information to decision makers. In the near future, these organizations could operate more along the lines of the FAO, i.e., as active repositories and synthesis centers of global-scale information (D. P. Turner, 2011). However, in practice there are continuous struggles to maintain institutional momentum with respect to global monitoring in the face of varying levels of national support. The Group on Earth Observations (GEO, 2017) is an international collaborative effort among governmental and nongovernmental organizations to develop a coordinated global monitoring system. Though recently completing an initial 10-year implementation plan, its operational component—the Global Earth Observing System of Systems (GEOSS)—has struggled to deliver useful data to stakeholders.

THE ROLE OF SATELLITE REMOTE SENSING

Global environmental monitoring got started with assemblages of point-based measurements at a sample of sites around the world. Data from meteorological stations, ships, and ocean buoys were collected and collated to formulate the first global averages and maps for climate variables. More recently, formal networks of measurements sites have been organized, e.g., the World Meteorological Organization (WMO) for climate data, ARGO (2017) for ocean data, the U.S. National Oceanic and Atmospheric Administration (NOAA, 2017a) for atmospheric observations, and FLUXNET (2017) for observations of land surface carbon fluxes.

Beginning around 1960, satellite remote sensing began to be used to create continuous surfaces (maps) of observed variables. Early weather satellites simply recorded the location of clouds, but soon a wide variety of satellite-borne sensors were in continuous operation. There are two basic types of satellite: (1) polar orbiting, and (2) geosynchronous. The polar orbiting satellites (which orbit Earth in about 90 minutes) generally record information from swaths of Earth's surface; the data are then patched together to create continuous fields. Geosynchronous satellites sit over a fixed point on Earth and transmit observations in real time (e.g., for clouds). Another important distinction in remote sensing is between passive sensors, which detect upwelling radiation (either from reflecting sunlight or emitted by the surface) and active sensors, which send a pulse of radiation to the surface and record the characteristic of the reflected radiation.

When the U.S. National Aeronautics and Space Administration (NASA) was created in 1958, it had no role in Earth science. However, missions to Mars and Venus in the 1960s raised the question of why those planets were so different from Earth. In addition, there was Cold War-era interest in mapping crop extent and productivity in Russia and elsewhere. With impetus from the Department of the Interior, NASA began development of the Landsat series of satellites in 1964 and initiated funding for university research scientists to explore the potential for land monitoring from space (Lauer, Morain, and Salomonson, 1997). NOAA was charged with space-based monitoring of weather and climate, and began continuous coverage of weather with polar-orbiting satellites in 1978. The NOAA Advanced Very High Resolution Radiometer (AVHRR),with an approximately 1 kilometer spatial resolution, proved to be useful for global observations of land cover and net primary production, as well as weather (Hastings and Emery, 1992).

By the 1990s, it was recognized by the scientific community that a new level of global-scale environmental monitoring was needed. General circulation models of the climate were suggesting that continued release of greenhouse gases would significantly warm the planet. The global science community clearly needed more and better information about the climate, the atmosphere, the land, and the ocean, and how they were changing. Inspired in part by the hot summer of 1988 and the dramatic testimony of climate scientist James Hansen that climate was already changing, the U.S. Congress initiated long-term funding for NASA's Earth Observing System (EOS) in the early 1990s. EOS was designed to make space-based observations relevant to understanding climate change. Monitored variables included temperature, precipitation, aerosols, ozone, snow/ice extent, land cover, and net primary production on land and in the ocean.

EOS was exemplary in many respects. From its beginning, there was a serious commitment to building a community of engineers, technicians, and researchers to maximize the scientific benefit of the program. All EOS observational data were made freely available from a network of data centers that was established to archive the complete record for each sensor (EOSDIS, 2017). Calibration (Datla et al., 2014) and validation (Justice et al., 2000) efforts were given high priority. The reward has been an extraordinary contribution to understanding the Earth system. NASA continues to develop, launch, and test a variety of Earth-observing sensors.

Other national space programs have also had strong satellite-based research programs, notably Japan and the European Union. The Committee on Earth Observation Satellites (CEOS, 2017) is an international coordinating institution that attempts to maximize cooperation and minimize redundancy among the national space agencies.

INTEGRATION OF GROUND-BASED AND
SATELLITE-BASED OBSERVATIONS

A significant issue with respect to managing and applying data from global monitoring efforts is integration of ground-based and satellite-based observations. Earth system scientists are often interested in both (1) how much of something is present, and (2) the rates of processes that determine what is there. Measurements on the ground give detailed information for particular sites, either at a point in time or over an interval if there is monitoring. These measurements for many sites are often assembled into networks, thus sampling a range of spatial and temporal heterogeneity. Remote sensing then offers the opportunity to move to the global scale.

Monitoring what is there (e.g., land cover distribution, vegetation carbon stocks, or polar sea ice extent) is somewhat more straightforward than monitoring process rates. The challenge in monitoring what is there lies in converting digital data on reflected or emitted electromagnetic radiation into a quantitative estimate. Different land cover types have characteristic spectral signatures: Ice reflects visible wavelengths much better than water does, and a warmer surface emits more infrared radiation than a cold surface. Given high-quality reflectance data from a satellite-borne sensor, and ground truth in the form of well-georeferenced ground measurements, some form of mathematical relationship can often be established between them that allows for mapping the property of interest based on the satellite imagery. Most satellites have multiple year lifetimes, giving the possibility to monitor changes.

To obtain the maximum benefit from Earth system observations, it is desirable to use them in building, parameterizing, validating, and applying simulation models that track the underlying processes. As noted in chapter 6, a simulation model is basically a series of mathematical relationships or equations (algorithms) that account for a process, e.g., photosynthesis. The algorithms might be purely empirical (e.g., photosynthesis decreases as the air temperature approaches $0°C$) or based on a more mechanistic approach (e.g., light of a certain wavelength induces certain biochemical events relevant to photosynthesis). Models for processes such as photosynthesis and evapotranspiration can be developed from ground-based observations. Then, with access to satellite data to drive them, the models may also be run in a "spatially distributed" mode, i.e., in each cell over a grid covering Earth's surface.

Satellite observations are integrated into the modeling process in several ways. Often, they help establish the initial conditions of global models, e.g., the distribution of vegetation types on the land surface as needed in global climate models. They can also provide model inputs, as in the information about vegetation greenness, which helps model processes such as photosynthesis. Satellite-based observations sometimes provide a means of validating model outputs; e.g., observed land temperature is used to evaluate simulated land surface temperature in a general circulation model, or satellite-based leaf biomass is compared to simulated leaf biomass (Luo et al., 2012). Satellite-based observations of clouds are particularly useful given the inherent problems with simulating clouds in climate models (Zhang et al., 2005).

The beauty of a spatially distributed simulation model is that it integrates information from a wide variety of inputs and produces estimates for process rates over large spatial domains that are not obvious from these inputs. Simulation models are critical tools in the "scaling" (i.e., extrapolating over space and time) of ecosystem process such as net primary production (D. P. Turner, Ollinger, and Kimball, 2004).

Recurrent research issues in Earth system monitoring concern the optimal spatial, temporal, and spectral resolution of the observations; e.g., for global monitoring of land cover change, a resolution of 250 meters or finer is required (Townshend and Justice 1988). Technical limits in sensor characteristics, data storage, and computation capacity are always a consideration, though they have increasingly become less relevant as technology advances. Generally, the engineers race ahead to build ever more refined observational capabilities, while climatologists and ecologist follow behind trying to understand what all the new raw data are actually telling us.

CLIMATE

The WMO collects meteorological station data (e.g., temperature, precipitation, humidity) from thousands of measurement sites distributed widely around the surface of the planet (WMO, 2017). More specialized networks gather data on specific climate variables, e.g., solar radiation (SURFRAD, 2014). The distribution of measurement stations is, of course, only a sample of the complete range of climate variability on the planet, and remote sensing bring us a big step closer to complete coverage. Satellites can systematically monitor surface temperature and precipitation as well as track the movement of clouds and development of storms.

Temperature measurement from space is based on the same concept of black body radiation we encountered in chapter 2, i.e., the principle that the temperature of a surface determines the characteristics of the electromagnetic waves it emits. This principle permits estimation of surface temperature based on infrared radiation emitted by the surface. Alternative approaches have also been developed, e.g., using the intensity of upwelling microwave radiation. Initially, there was a discrepancy between the historical temperature record from ground stations and that from satellite observations. However, the two data sources are largely reconciled now, with both showing an increasing trend in global mean temperature (Wigley et al., 2006).

Precipitation monitoring is based on retrieval of reflected radar signals and upwelling microwave radiation (Ceccato and Dinku, 2016). The Global Precipitation Measurement Mission (GPMM, 2017) is a joint effort by NASA and the Japan Aerospace Exploration Agency that integrates observations from a variety of sensors.

Storms are observed with many sensors, both geosynchronous and polar orbiting. The NOAA GOES (Geostationary Operational Environmental Satellite) series began in the 1990s, and has continued as the principal observation satellite for weather monitoring (GOES, 2017). As noted, NOAA has also supported the AVHRR series of polar-orbiting weather satellites, beginning in the 1980s.

Reconstruction of the historical climate is based on assimilation of observations into climate models (Saha et al., 2010). In this approach, the climate model (see figure 6.1) is regularly updated with a massive amount of observational data from the ground and satellites, thus recreating the climate as it was observed. This approach permits comprehensive assessment of ongoing changes and trends based on observations. Assimilation of satellite data into climate models also plays a role in standard weather forecasting efforts.

The Earth system science climate research community collaborates to produce an annual State of the Climate report that comes out each June (SOTC, 2016). It describes global and regional climate for a given year and places it in the context of the historical record.

CRYOSPHERE

The term *cryosphere* refers to the global distribution of snow and ice. Satellite remote sensing is particularly effective in monitoring the geographic distribution of snow and ice because of the high albedo (i.e., reflectance) difference between snow/ice and land/water.

The distribution of sea ice is monitored by satellite-borne sensors such as MODIS (Moderate Resolution Imaging Spectroradiometer), and is of special interest because in recent decades Arctic sea ice extent and thickness has decreased much faster than predicted by climate models. At the same time, Antarctic sea ice has slightly increased. The National Snow and Ice Data Center (NSDIC, 2017) continuously updates the satellite record (figure 9.1).

The reduction of snow cover in the northern hemisphere is also a harbinger of climate warming. The retreat of glaciers in the temperate and tropical zones is well documented by both ground and satellite-based observations (Paul, 2014). Specialized radar sensors detect frozen ground, which is also expected to retreat in coming decades. NASA's SMAP (Surface Moisture Active Passive) sensor, launched in 2014, monitors surface soil moisture and frozen ground, albeit at a lower resolution than originally planned because of partial instrument failure (SMAP, 2017).

We know that the volume of snow and ice in the Greenland ice cap is sufficient to raise sea level by about 7 meters if it all melted (IPCC, 2014e). Monitoring ice volume is not possible with optical sensors, but can be done with active radar sensors. The return time of a pulse of electromagnetic radiation gives an indication of ice surface height; changes in height and extent of the ice caps thus reveal changes in their volumes. The NASA ICESat1 (Ice, Cloud, and Land Elevation Satellite) sensor established the feasibility of ice cap monitoring but is now out of service. A successor is under development and scheduled for launch in 2017. That kind of service

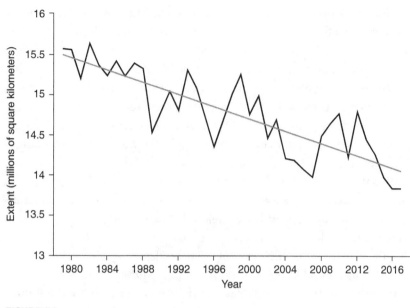

FIGURE 9.1

Trend in average area for Arctic sea ice in May. Source: National Snow and Ice Data Center.

gap is an unfortunate reality for satellite-borne sensors because of their huge costs, long lead times, and the political nature of their funding.

HYDROSPHERE

Our focus here is freshwater monitoring; ocean monitoring is treated separately below. Much of humanity resides along streams and rivers. Thus, the volume and seasonality of stream flow is one of the ecosystem variables with the longest record of observation. The Palermo Stone in Egypt, dating to about 2500 BCE, lists the annual maximum height of the Nile River in that era. In recent decades, national networks of stream gages have grown. These observations are beginning to be studied systematically (GTN-R, 2017) but are not routinely shared and aggregated to the global scale.

The availability of real-time satellite-based information on storms, drought, and flooding helps in disaster management. Monitoring information that reveals trends in precipitation, stream flow, and groundwater will help with planning and adaptation to climate change.

The most compelling recent advance in monitoring the hydrosphere has been results from the GRACE sensor. The Gravity Recovery and Climate Experiment (GRACE, 2017) was launched in 2002 and reports monthly the total water content of grid cells on the order of 500 kilometers on a side. These observations are based on subtle changes in the gravitational field, and the measurements are capable of sensing changes in aboveground (e.g., glaciers and reservoirs) as well as belowground (aquifer) water storage. As with all satellite-borne sensors, GRACE circumvents national-level restrictions on information flow. This capability is highly relevant as the incidence of cross-border water disputes increases. GRACE has revealed that several important aquifers (notably the Ogallala Aquifer in the United States and the aquifer underlying the Punjab region in India) are being depleted (Famiglietti and Rodele, 2013). Like ICESat, GRACE was an experimental mission that is now over. Fortunately, a new sensor is to be launched in 2017 with an objective of systematically monitoring groundwater changes.

ATMOSPHERE

Pollution of the global atmosphere became an issue starting after World War II. The rising level of radionucleotides from testing nuclear weapons provided an incentive for an international agreement to ban further aboveground testing (chapter 8). In the 1980s, the specter of "acid rain" was evoked by rising levels of nitrogen and sulfur emissions derived from burning of fossil fuels. Corresponding observations of rainfall and aquatic chemistry prompted various international agreements to limit emissions, usually with associated monitoring and reporting requirements. As we have noted (chapter 1), Charles David Keeling devised an instrument capable of accurately measuring atmospheric CO_2 concentration around 1955. His measurements at the Mauna Loa volcano in Hawaii helped alert the world to the magnitude of anthropogenic impacts on the global environment.

Besides CO_2, the capability to measure methane, ozone, and nitrous oxide in the atmosphere is also now well established and the historic records of their concentrations at many stations are readily available (NOAA, 2017a). Concentrations of CO_2 and methane are higher in the northern hemisphere than the southern hemisphere, suggesting larger sources in the more industrialized north. The recent record of methane (figure 9.2) has atmospheric scientists puzzled because of the approximately 5-year hiatus in its concentration growth around the year 2000.

Despite vigorous vertical mixing and the general circulation of the atmosphere, there is considerable spatial and temporal heterogeneity in its chemical composition. That heterogeneity in concentration is related to heterogeneity in the source and sinks of particular compounds and patterns of atmospheric circulation. Indeed, information on the spatial and temporal patterns in atmospheric chemical composition can be used to infer the magnitude of sources and sinks.

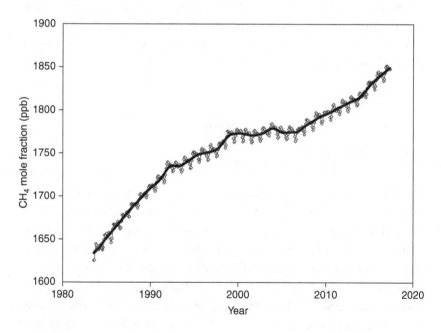

FIGURE 9.2

Record of methane concentration in the atmosphere at Mauna Loa. Courtesy of Ed Dlugokencky, NOAA/ESRL.

In what is known as an "inversion" approach, sources and sinks of CO_2 are inferred from the spatial and temporal patterns in measured concentrations, which are combined with models of atmospheric circulation (Peters et al., 2007). CarbonTracker is the inversion modeling setup used by NOAA scientists to map CO_2 sources and sinks globally; it delivers an annual update (CarbonTracker, 2017). In support of inversion modeling, NOAA has established a global network of sites where flask samples of the atmosphere are collected on a regular basis. Concentrations of all the major trace gases as well as various isotopic forms are determined in these samples.

Concentrations of trace gases in a column of air can be estimated by upward-looking ground sensors, and by airborne or satellite-borne sensors. A satellite-borne methane sensor recently detected anomalously high methane concentrations over the Four Corners area in the southwestern United States, probably related to leakage from fossil fuel mining there (Kort et al., 2014). Satellite-borne sensors now estimate the abundance of CO_2 in columns of air measuring 100 kilometers on a side (OCO, 2017). These data sets and associated modeling efforts to map carbon sources and sinks will provide a common reference point as international climate change mitigation negotiations proceed.

LAND

Land Cover and Land Use

We noted in chapter 4 that approximately half of Earth's land surface has been converted from primary vegetation to human-influenced vegetation. This change is ongoing and has significant implications for availability of farmland, conservation of biodiversity, surface energy balance, and greenhouse gas emissions. Our ability to monitor global land cover began with the launch of the initial Landsat mission by NASA (Cohen and Goward, 2004). Subsequently, a continuous series of Landsat satellites has been placed in orbit; the value of the associated imagery time series (over more than 40 years) is well recognized. The Landsat-borne sensors are multispectral: They detect reflectances of solar radiation in multiple wavelength bands (visible and near infrared) and (on recent versions) upwelling radiance in the thermal infrared region of the electromagnetic spectrum (Ceccato and Dinku, 2016). The spatial resolution of the imager (i.e., the pixel size) in the recent Landsat sensors is about 30 meters. This resolution is fine enough to readily capture most processes of land cover change, particularly urbanization and deforestation. The Landsat temporal resolution is global coverage every 16 days. The digital data from the sensors is downloaded to an international ground station network, processed by U.S. Geological Survey scientists, and made freely available to the global science community (Woodcock et al., 2008).

Vegetation strongly absorbs red and blue wavelengths (used in photosynthesis) and reflects near-infrared wavelengths. Thus, various ratios of the reflectances have been developed to monitor vegetation, notably the Normalized Difference Vegetation Index (NDVI) or greenness index. Examination of 28 years of Landsat NDVI at high latitudes found increases in greenness over 29 percent of the studied area, likely related to climate warming (Ju and Masek, 2016). Landsat is especially effective in change detection and thus has been a key tool in the effort to track deforestation (M. C. Hansen et al., 2013). It also is used to track forest disturbances, including harvest, fire, and insect outbreaks (Kennedy, Yang, and Cohen, 2010), and to map forest stand age (Cohen, Spies, and Fiorella, 1995).

The global record of Landsat imagery since 1985 has recently been assembled for the purpose of monitoring global deforestation. Global Forest Watch (2017), a nongovernmental organization (NGO), processed the data to emphasize changes in forest cover. It now makes a regularly updated dataset available to nonspecialist users. This approach has served to broaden the range of individuals and organizations involved in the assessment and reduction of deforestation.

Carbon Flux on Land

We've empathized that a key role of the biosphere in the metabolism of the Earth system is in capturing solar energy and producing biomass. This process, referred to as net primary production (NPP) and quantified as grams of biomass produced per meter squared (m^2) per year, is measured most simply by weighing biomass, e.g., harvesting the complete corn plants at the end of the growing season for a given area. NPP measurements in non-crop ecosystems is more complicated than in cornfields (Gower, Kucharik, and Norman, 1999) but have now been made throughout the terrestrial biosphere.

Spatial and temporal patterns in NPP are of great practical interest because (1) NPP is the base of the food chain in most ecosystems, (2) it has a significant role in the global carbon cycle, and (3) it contributes to production of food and fiber for human use. Compilations of NPP measurements and knowledge of the global area of each ecosystem type suggest a global NPP (expressed in terms of carbon, which constitutes about half of biomass) of about 60 petagrams of carbon (PgC) per year (Saugier, Roy, and Mooney, 2001). (Recall that $1 \, Pg = 10^{15}$ g, or 1,000,000,000,000,000 grams.)

The other terrestrial biosphere carbon flux variable of intense interest to Earth system scientists is the net ecosystem exchange (NEE) of carbon (expressed as grams of carbon per m^2 per unit time). This flux is primarily the net effect of carbon uptake from the atmosphere by way of photosynthesis (i.e., gross primary production) and carbon returned to the atmosphere by way of autotrophic and heterotrophic respiration. NEE is measured at eddy covariance flux towers based on concentration differences in the downward and upward components of eddies in the atmosphere (Baldocchi, 2003). Long-term monitoring of NEE was first undertaken at Harvard Forest in the northeastern United States (Wofsy et al., 1993), and a global network of flux towers (FLUXNET, 2017) now continuously monitors NEE in all the major terrestrial ecosystems. These flux towers are essentially sentinels keeping watch on the metabolism of the terrestrial biosphere.

The surge of satellite data available in recent decades has made global monitoring of terrestrial NPP and NEE feasible (Schimel et al., 2015). My own research involved evaluating what the MODIS (Moderate Resolution Imaging Spectrometer) sensor can tell us about vegetation productivity (D. P. Turner et al., 2005). MODIS is one of the NASA EOS sensors and was put into orbit in 2000 (MODIS, 2017). The spatial resolution of the sensor is about 1 kilometer, and it acquires images in several wavebands for the whole Earth surface once a day. These data are transmitted to a processing facility at NASA, where they are cleaned up and packaged for delivery to users. As with Landsat data, MODIS data are freely available to the global research community over the Internet in near real time (LP DAAC, 2017). The MODIS Science Team is responsible for producing an annual global NPP product (Running et al. 2004).

Net primary production is derived from gross primary production (GPP) by subtracting autotrophic respiration (approximately 50 percent of GPP). The approach to estimating GPP from satellite imagery is commonly based on a light use efficiency (LUE) value, i.e., the amount of carbon fixed per unit of sunlight absorbed by the vegetation canopy (Running et al., 2004). LUE is routinely estimated from NEE measurements at eddy covariance flux towers (D. P. Turner et al., 2003). LUE differs by vegetation type; e.g., growth in boreal forest is less per unit sunlight absorbed than in temperate forests. LUE is also reduced by factors such as temperature extremes and low relative humidity.

To derive GPP globally, MODIS imagery gives information on the vegetation cover type and the amount of leaf biomass, which determines how much solar radiation is absorbed. Daily weather from reanalysis of meteorological station data (e.g., DAYMET, 2017) provides spatially distributed information on temperature, humidity, and solar radiation. This set of information for each grid cell is run through the MODIS LUE-based GPP algorithm to estimate daily GPP. More recent satellite-borne sensors such as the Orbital Carbon Observatory (OCO, 2017) also estimate GPP but rely on emissions from vegetation of a very narrow band of electromagnetic radiation related to photosynthetic activity (Frankenberg et al., 2014).

As part of the MODIS sensor validation effort, the MODIS GPP and NPP products have been evaluated at field sites ranging from the Arctic tundra at Barrow, Alaska, to a tropical rain forest site in the middle of the Amazon Basin (D. P. Turner et al., 2006). These comparisons of satellite-based NPP estimates and ground measurements were also used to refine the MODIS NPP algorithm. Analysis of the time series of MODIS-based global NPP estimates (2000–2009) suggests a possible decline, driven largely by drought in tropical forests (Zhao and Running 2010).

Other land applications of MODIS data include monitoring effects of drought on vegetation, e.g. the intense heat wave and drought that hit Europe in 2003 (Reichstein et al., 2006), as well as monitoring forest fires and more benign phenomenon such as the green wave that moves north on land in the northern hemisphere spring. The "green marble" shows the annual sum (or mean) of a vegetation index (an indicator of summed photosynthesis) for each high-resolution grid cell on land (Graham, 2013).

Other Land Variables

MODIS is just one of a wide array of sensors now orbiting the Earth and delivering terabytes (10^{12}) per day of observational data to receiving stations on the ground. Other sensors operate at much higher spatial resolution, essentially able to resolve individual trees. Higher spectral resolution sensors are also available and informative with respect to variables such as foliar chemical properties (Asner and Martin, 2009). Active sensors, e.g., radar and lidar, send pulses of electromagnetic radiation to the surface and measure the characteristics of the reflected light. Active sensors have been used for global mapping of forest height and biomass (Lefsky, 2010; Saatchi et al., 2011). Eventually, this technology may deliver an annual high spatial resolution report of global biomass based on satellite-borne sensors. This product would have tremendous application in verifying national-level commitments to reduce carbon emissions by slowing deforestation or increasing forest carbon sequestration.

Biodiversity

Biodiversity is a special case with regard to monitoring because even detection of presence or absence of plant and animal species can be difficult. Information about population structure of many species is also urgently needed but rarely available (Ripple et al., 2016). Remote sensing can be helpful at the level of monitoring vegetation cover type and functional diversity (Jetz et al., 2016), but generally not at the species level.

The most advanced effort for collecting and disseminating global-scale information about biodiversity is the periodic release of a list of threatened species by the IUCN (16,119 species of animal and plant were listed as of 2000). This NGO was founded in 1948 and, besides research, is actively involved in development of national and global strategies for conservation of biodiversity. Within the GEOSS framework, biodiversity is one of the more well-developed areas of benefit to Society, but the biodiversity monitoring network is still in the design phase (Scholes et al., 2012).

OCEAN

The global ocean provides many important ecosystem services and thus requires an international system of governance and monitoring. Notably, the ocean is currently a sink for approximately 25 percent of fossil fuel carbon emissions (GCP, 2017) and is the primary source of protein for 2.6 billion people (FAO, 2014).

The ocean is monitored by a combination of satellites, floating and submerged sensors, and ship cruises. As with climate observations, there is good international coordination with respect to observations of physical properties through the Global Ocean Observing System (GOOS, 2017). Physical/chemical variables of interest include sea level, sea surface temperature, CO_2 concentration, and circulation patterns. Biological variables that are monitored include phytoplankton production, coral reef health, and fish/whale populations.

Monitoring of sea level has established an approximately 3 millimeter per year rise in recent decades (IPCC, 2014e). In this case, there has been international coordination (GLOSS, 2017) to combine results from a network of coastal observing stations and multiple satellite altimetry missions. The observed sea level rise helps validate models of ice cap melting and ocean thermal expansion.

Sea surface temperature is also tracked by satellites. Results are particularly useful in understanding and predicting large storms and inter-annual variation associated with ocean circulation features like the El Niño–southern oscillation. However, satellites only measure surface conditions. Recently a network of free-drifting profiling floats has begun to produce detailed information on depth profiles for temperature, as well as on rates of ocean circulation (ARGO, 2017).

Ocean NPP is monitored by a MODIS sensor based on estimates of chlorophyll concentration. The magnitude of ocean NPP is approximately the same as terrestrial NPP (Field, Behrenfeld, Randerson, and Falkowski, 1998). However, the uncertainty is higher, and it remains controversial whether there is a trend of decreasing phytoplankton production (Behrenfeld et al., 2016). As on land, much of NPP in the ocean is rapidly cycled back into CO_2 by way of decomposition. A small proportion sinks to the ocean floor, which means longer term sequestration. Significant efforts are in place to track that proportion (Siegel et al., 2014).

The topping out of the global fish catch (discussed in chapter 5) has alerted the world to the need for better monitoring and regulation of commercial high seas fisheries as 64 percent of the ocean is outside national jurisdictions. At the global scale, Global Fishing Watch (2017) now tracks large vessels using the satellite-accessible automatic ship identification (AIS) information. Ship locations and behavior are juxtaposed with the boundaries of the increasing number of marine protected areas to reveal fishing vessels in illegal locations. This is an emerging capability that will permit improved assessment of the effectiveness of globally agreed upon marine regulatory policies (McCauley et al., 2016).

RECENT DEVELOPMENTS IN EARTH OBSERVATION

Throughout this chapter, we've noted the many self-organized networks for observations of specific variables. These networks have proven valuable in synthesis efforts that advance the relevant scientific disciplines, and for the assessments that drive changes in environmental policy. Participation in such networks is usually voluntary, with participants relying on their own national-level resources. This structure generally works, but is not conducive to stability. Hence, the drive for an institution, such as GEOSS, or an internationally supported World Environment Organization (see chapter 8) that would make the monitoring efforts more sustainable. Such an institution could provide consistent annual global summaries about Earth system structure and function for the public and for policy makers.

As humanity begins to truly engage in reducing emissions of CO_2 and methane, a reliable carbon cycle monitoring system will be especially important. The products of this system would be relatively high spatial and temporal resolution estimates of carbon pools and fluxes, with associated uncertainties, covering the global domain. The approach would rely on satellite-based, and ground-based observations as well as models. Multiple data streams could be assimilated by way of advanced model-data fusion techniques.

Implementation of a global carbon monitoring system will need to be a collaborative effort on an international scale. This is necessary both to satisfy the requirement for transparency (for any system deployed unilaterally will not enjoy widespread credibility), as well as the very practical need to pool resources in the form of physical infrastructure, and scientific and engineering expertise (Ciais et al., 2014). However, the institutional framework for an internationally funded carbon cycle monitoring system has not been worked out.

More generally, concerning the array of satellite-borne sensors, there is consistent progress in development of new sensors and their operational implementation (e.g., GRACE and ICESat). However, satellites have limited lifetimes and there is concern about possible gaps in coverage in future years. Remarkably, the two MODIS sensors from the EOS program are still operational (more than 15 years into an expected mission lifetime of 5 years). A MODIS replacement is now in orbit (VIIRS, the Visible Infrared Radiometer Suite) and NASA is scheduled to launch the follow-on JPSS-1 satellite in November 2017. Optimally, there will be enough continuity across sensors that they can be intercalibrated to produce the kind of a continuous time series needed for trend analysis of variables such as global NPP.

The European Space Agency (ESA) has recently begun to launch a series of Earth observing sensors (ESA, 2017). The Sentinel 2 mission is similar to the Landsat mission, and ESA has agreed to share its data. The combination of Sentinel 2 and Landsat imagery will help in the production of a time series of Landsat resolution imagery suitable for a variety of applications, including phenology studies. After many years of debate, the United States is now firmly committed to long-term continuity of the Landsat mission. The data-sharing agreement covering Sentinel 2 and Landsat points toward the need for better integration of technical standards and data sharing across satellite platforms (Wulder and Coops, 2014).

The Global Climate Observing System recently summarized the kind of atmosphere, ocean, and land observations need to understand the climate (GCOS, 2017b). A proposal for a Carbon Cycle Observing System has likewise been put forth (Ciais et al., 2014). For many of the variables cited in these proposals, operational observing systems are in place (NCEI, 2016). Science-funding agencies seem to have finally overcome a historical reluctance to fund monitoring projects (C. D. Keeling, 1998). However, a tremendous amount of international coordination is still needed to support the kind of information flow that will be required to begin actively managing the Earth system.

ESSENTIAL POINTS

As is evident from this survey of current global-scale monitoring efforts, the nascent global monitoring community is still in the early stages of developing a comprehensive global monitoring system. Networks of ground-based stations, as well as complimentary satellite-based observations, are needed. Advocates for a revamped UNEP, or a new World Environmental Organization, note that monitoring capacity is clearly required as a core component of a broader global environmental governance system.

IMPLICATIONS

1. Global monitoring is an integral part of the effort to understand, forecast, and manage Earth system structure and function. It requires national-level commitments to basic and applied research as well as international coordination to maximize the return on societal investment.

2. Satellite-based monitoring helps diminish the information restrictions imposed by traditional national boundaries, thus supporting transparent international negotiations to address global environmental change issues.

3. A comprehensive global monitoring program includes site-level and satellite-based observations but also involves modeling the relevant processes, thus providing a basis for steadily improving our understanding of the underlying physical and biophysical mechanisms.

4. An internationally coordinated effort (and institution) is needed to annually summarize the currently disparate, mostly research-based, results of global monitoring.

FURTHER READING

Ciais, P., Dolman, A. J, Bombelli, A., Duren, R., Peregon, A., Rayner, P. J, . . . Zehner, C. (2014). Current systematic carbon-cycle observations and the need for implementing a policy-relevant carbon observing system. *Biogeosciences, 11*, 3547–3602.

Hansen, M. C, Potapov, P. V, Moore, R., Hancher, M., Turubanova, S. A., Tyukavina, A., . . . Townshend, J. R. (2013). High-resolution global maps of 21st-century forest cover change. *Science, 342*, 850–853.

10

INTEGRATING SOCIAL AND ECOLOGICAL SYSTEMS

Global change research must reorient itself from a focus on bio-physically oriented, global scale analysis of humanity's negative impact on the Earth system to consider the needs of decision makers from household to global scales.

—R. S. DeFries et al. (2012)

THE SYSTEMS WORLDVIEW

The business-as-usual scenario is generally recognized as leading to a degraded biosphere, and potentially to catastrophe for advanced techno-logical civilization (see chapter 6). Hence, the aspiration for a "Great Transition" (as per the Anthropocene narrative), i.e., a transition to global sustainability (Gell-Mann 2010; NRC, 1999) and integration of techno-sphere and biosphere (Folke et al., 2011). In this book, we have mostly taken a diagnostic approach and analyzed global-scale structures and processes. The global scale helps frame the issues, and global-scale envi-ronmental governance activities help establish a shared understanding of the problems. But to actually begin addressing the issues raised by global-scale analyses, environmental governance must often be operationalized over smaller domains (DeFries et al., 2012) and successes there scaled up to the global domain (Lubchenco, Cerny-Chipman, Rehmer, and Levin, 2016). This chapter will examine how coupling of social and ecological systems at multiple spatial scales could contribute to achieving a Great Transition.

Ecological systems (i.e., ecosystems) were, until recent human interventions, the product of biological evolution acting over billions of years. An ecosystem consists of biotic communities, the abiotic environment, and associated biogeochemical cycling processes (see chapter 3). An ecosystem can be characterized by its natural capital, i.e., the stocks of living matter, and the structured ecological relationships among its organisms (Daily, 1997). In addition, an ecosystem has measurable flows of energy and mass by which it is maintained. The prehuman Gaian planet was an array of ecosystems, and the biosphere itself, along with its abiotic environment, could be considered the global ecosystem.

Human social systems have evolved more recently, and mostly by way of cultural evolution rather than biological evolution. A society is based on a network of interpersonal relationships and a shared culture (i.e., language, customs, and beliefs). A society has social capital in the sense of members who function well together and share traditions that help maintain order and continuity. With the development of technology and the invention of money, human societies have expanded and become territorial. They are now also associated with technosphere capital (e.g., buildings, transportation infrastructure, communications infrastructure) and financial capital (the power to purchase goods and services and to invest).

Social systems interface with ecological systems by way of management (or mismanagement) of natural resources. A hunter-gatherer tribal society harvests plants and animals, thus impacting the population structure of the harvested organisms. The use of fire by hunter-gatherers to maintain an open environment that supports grazers is well documented (Lightfoot and Cuthrell, 2015). Once humans developed agriculture, the coupling of social and ecological systems tightened. In contemporary agricultural ecosystems, productivity is increased by management-related subsidies of water and nutrients; then much of the plant and animal production is extracted for human use.

Societies strive to manage their associated natural resources to maintain a reliable flow of ecosystem services (Daily and Matson, 2008). That requires a good understanding of the ecosystem stocks and flows, as well as comprehending restraints on the rate of harvest. Periodically in human history, societies have degraded their natural capital (e.g., overgrazing), lost the natural resource flows upon which their society depended, and collapsed (Diamond, 2005).

Early in human history, mismanagement of natural resources was punished by reduced productivity—a clear negative feedback. However, as the capabilities of the technosphere evolved, it became possible for human social units to survive and grow by consuming natural capital at one location, thus potentially ending the flows, but escaping the cost to themselves by moving on to another unexploited location. The story of the initial wave of timber harvesting in the United States follows that pattern, moving from the forests in the Northeast, to those in the Midwest, and on to those in the Pacific Northwest (M. Williams, 1989). The global fishery has been similarly exploited (see chapter 5). The associated losses of natural capital largely did not register with the resource harvesters; they simply moved on. Other strategies, such as intensification of exploitation with improving technologies, have also served to insulate human societies from the near-term consequences of resource degradation (Raudsepp-Hearne et al., 2010b).

This mode of natural resource exploitation continues in many locations around the planet. However, as we begin to reach the planetary boundaries (described in chapter 8), fewer and fewer unprotected resources remain. To continue flows of ecosystem services, the choices come down to (1) spend more financial capital to expand technosphere infrastructure for capturing dwindling stocks, or (2) devise sustainable natural resource management systems (Matson, 2009). Hence, the origin (Kates et al., 2001) and rapid evolution (Leemans 2016) of the discipline of sustainability science (Kates, 2010; Sachs, 2015).

We are rapidly learning that it is possible to manage natural resources in a way that maintains natural capital and yields a harvestable surplus for human consumption. The task now is to develop and apply sustainable natural resource management schemes at all geographic scales, including the global scale.

MANAGEMENT OF SHARED RESOURCES

A core issue in advancing the sustainability paradigm is how to collectively manage natural resources over large geographic domains. An early formulation of the difficulties with managing a natural resource of a constrained size, which is available to multiple users and diminished by excess use, was Garret Hardin's "tragedy of the commons" (Hardin, 1968). In that model, the commons is a public resource such as communal grazing land. The tragedy is that the benefit of a shepherd adding another sheep to his flock accrues to the shepherd, but the cost in terms of damage to the communal land from overgrazing is paid by the community. Thus, shepherds will continue to add sheep even after the carrying capacity has been reached, and the ecosystem will continue to deteriorate from overgrazing.

A similar dilemma arises in the case of open-access resources. Open seas fisheries are traditionally available to all users, and correspondingly have frequently become overfished and depleted (Ostrom, 2008). The global atmosphere is used as a dumping ground for carbon dioxide (CO_2) from fossil fuel combustion. Regional aquifers that cross international boundaries are tapped by users with no consideration of users elsewhere. These are cases of common pool resources (CPR), which are characterized by non-exclusivity (anyone has a right to use them), and divisibility (use by one party subtracts from total availability).

There is a related "tragedy of ecosystem services" that applies to privately owned resources (Lant, Ruhl, and Kraft, 2008). In this model, a land owner has less incentive to maintain a flow of ecosystem services that benefits the community than incentive to manage for maximizing return on investment. An intact forest may do a better job than an agricultural field in regulating stream flow, but the agricultural field may yield a better financial return. The general tragedy in natural resources management is often lack of institutions and social capital to collectively manage resources so as to avoid loss of natural capital.

The two typical governance approaches to managing a shared resource that is being abused are (1) for the community of users to impose a regulatory framework, or (2) to privatize the resource. Both approaches have strengths and limitations (Lueck, 1995; Ostrom, Burger, Field, Norgaard, and Plicansky, 1999). With the regulatory approach, an administration and enforcement infrastructure must be established and maintained to ensure rule of law. In addition, there is always the problem of free riders, i.e., users who do not contribute to maintenance of the resource. Thus, this approach may be impractical or have costs that exceed the value of the resource. Privatizing the resource may inspire the owner to protect it and to manage it in a way that conserves its productivity. However, this approach likely deprives some community members of access, and opens the path to the tragedy of ecosystem services.

In recent decades, environmental sociologists and political scientists have studied alternatives to CPR management that rely on various forms of property rights and ways of collaborating to self-regulate (Adams, Brockington, Dyson, and Vira, 2003; Dietz, Ostrom, and Stern, 2003; Ostrom, 1990). Local management of the lobster fishery in Maine is a paradigmatic example of success over a small geographic domain in the United States (Acheson, 2003; J. Wilson, Yan, and Wilson, 2007). There, coordination among various institutions, and agreements among lobster catchers about geographic boundaries of harvest rights and about optimal catch limits, has produced a sustainable resource. What works to build a sustainable management scheme varies depending on the environmental setting and the people involved, but generalizations have begun to emerge (Ostrom, 2009).

THE SOCIOECOLOGICAL SYSTEM CONCEPT

The socioecological system (SES) is a core concept in the theory of CPR management (Berkes, Colding, and Folke, 2003; Ostrom, 2007). An SES can be viewed as "a complex adaptive system consisting of a biogeophysical unit and its associated actors and institutions" (Glaser, Ratter, Krause, and Welp, 2012, p. 4). Geographic boundaries, as well as the relevant ecosystem services and identified managers, help define an SES. As with the concept of an ecosystem, an SES is considered a self-organizing entity.

A network of interactions and feedbacks link the biophysical and social components of an SES. Communities of stakeholders jointly use, monitor, and conserve the relevant natural resource. Key issues in development of SES theory have been (1) how to build cooperative relationships within an SES, (2) how to build resiliency into an SES, and (3) how to coordinate across SESs that operate at different scales or are focused on multiple resources, or both.

Understanding an SES requires understanding of the social system (including political, economic, and power dynamics), the ecological (bio-physical) system, and the interactions between them. Hence the need for transdisciplinary research involving social scientists and ecologists.

Some of the features of successful SESs are as follows (Ostrom, 2009).

1. A buildup of social capital. Face-to-face group meetings of stake-holders in which opposing viewpoints are aired is a start. Relationships of trust are more likely if group members share a geographic domain and cultural norms.

2. Developing a shared understanding of the functioning of the resource, e.g., productivity, spatial distribution, interactions with other resources, and interannual variation. This understanding can be based on a capacity to simulate (model) the resource and perform sensitivity analyses, including its response to environmental variability.

3. Ensuring that the group has the power to establish and enforce agreed-upon rules for resource use. A sense of agency both in terms of the group as a whole and individuals within it is critical (Kenward et al., 2011; Westley et al., 2013).

4. Acquiring capacity to monitor the resource. Monitoring is necessary for adaptive management and for group evaluation of enforcement effectiveness.

5. Aiming for consensus rather than imposing majority rule. The losers in a majority rule decision may sabotage the management scheme.

6. Maintaining an adaptive management approach (Folke, Hahn, Olsson, and Norberg, 2005). Ecosystems are complex, and understanding of the biophysical system is often limited. Thus, learning by doing may be the only viable approach (DeFries and Nagendra, 2017).

The description of dynamics for both the biophysical and social sub-systems within an SES can be cast in terms of adaptive cycles. This formulation takes account of the ubiquity of disturbance and recovery. The adaptive cycle includes four phases: exploitation, conservation, release, and reorganization (Gunderson and Holling, 2002; figure 10.1). It can be readily illustrated with respect to a biophysical system using the case of forest succession. A young growing forest (exploitation phase) builds natural capital in the form of biomass and biodiversity. In a mature state (conservation phase) the forest stand is characterized by tight coupling of producer, consumer, and decomposer organisms such that nutrient leakage is low and overall stability is high. Eventually, a disturbance such as forest fire disrupts the connectedness within the forest ecosystem, and resources become available (release phase). The elevated level of energy (sunlight) and nutrients available in the release phase allows the establishment of a variety of organisms, especially rewarding fast-growing, weedy

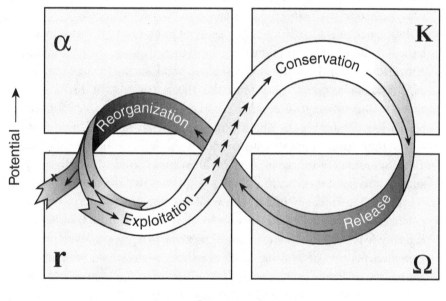

FIGURE 10.1

The adaptive cycle. The symbols r and K refer to species that are adapted to early or late stages of succession (Pianka, 1970); α (first letter of the Greek alphabet) refers to a beginning; Ω (last letter of the Greek alphabet) refers to an ending. The tail labeled x refers to the potential for the system to undergo a regime shift (Gunderson and Holling, 2002). Copyright © 2002 Island Press. Reproduced by permission of Island Press, Washington, D.C.

species (reorganization phase). As the ecosystem fills in and competition for resources increases, there may be selection for slower growing but longer lived species, and the exploitation phase plays out again.

There are potential analogues to these phases in the dynamics of social systems (Westley et al., 2013). Disturbances or shocks to social systems come in forms such as changes in leadership. Sitting powers are removed and reorganization becomes possible. New leaders, and innovative ideas, change institutional structures and modes of operation. Ultimately a stable period of effective governance is achieved. If the social system becomes rigid (overconnected) and ineffective, a new disturbance may be precipitated. Within an SES, an adaptive cycle may play out only in the biophysical subsystem, only in the social subsystem, or in both simultaneously.

Resilience refers to the ability of a system to absorb stress and disturbance without losing functionality (B. Walker, Holling, Carpenter, and Kinzig, 2004). Optimally, a system—biophysical, social, or socioecological—does not become so disorganized in the face of stress or disturbance that it begins to lose its natural or social capital and its capacity for self-regulation. Designing an SES to maximize resilience makes it more robust than specifying intricate top-down regulation (Lubchenco et al., 2016). Sources of resilience in biophysical systems include (1) biodiversity, such that relevant species are available for all phases of the adaptive cycle; (2) spatial heterogeneity or patchiness, along with connectivity between patches, so there is always refuge from disturbances; and (3) buffering capacity (e.g., high soil water-holding capacity in a forest ecosystem) so extended periods of unusual conditions can be tolerated. Analogous sources of resilience in SESs include (1) diversity in the human and biophysical components of the SES, (2) managing at multiple spatial scales, (3) applying an adaptive management approach (Anderies, Walker, and Kinzig, 2006; Matson, Clark, and Anderson, 2016), and (4) redundancy, so that loss of one component is not destabilizing. Managers can help build resilience in an SES by assessing its vulnerability (of both the social and biophysical subsystems) to stress and disturbance (B. L. Turner, 2010), possibly followed by adaptive interventions.

In the initial construction of an SES, a critical design feature is the nature of the feedbacks between the management and the biophysical components (box 10.1). Sustainability performance measures (e.g., conservation of biodiversity) provide a basis on the management side for evaluating particular management practices. The evaluation, of course, relies on adequate monitoring.

SESs have initially tended to be developed at rather small scales, in part because the possibilities for face-to-face dialogue, for monitoring, and for enforcement are enhanced. However, globalization increasingly influences

---

BOX 10.1
Getting Incentives Right

Incentives refer to things that motivate actors to do something—"the expected rewards and punishments that individuals perceive to be related to their actions and those of others" (Matson, Clark, and Anderson, 2016, p. 91). Incentives are more likely to be effective if "they reinforce behaviors that enable individual actors to act in their self-interest in a fashion that aligns their behavior with the larger goals of the communities or society" (Lubchenco et al., 2016, p. 14508). Management must often target decision making by individuals directly involved in resource utilization because the cumulative effects of many actors controls the dynamics of the socioecological system in which they are embedded.

Secure access fisheries, in which a sustainable level of harvest is divided among an authorized set of harvesters, succeed by aligning economic incentives of harvesters with broader conservations goals. Certification programs (see chapter 8) likewise foster restraint among producers because it pays off in terms of consumer interest. Reputation-based incentives operate through more subtle motivations, such as personal or national reputation for environmental responsibility.

Given a well-designed incentive structure, and monitoring of the resource, strong positive feedbacks can develop between environmental quality and motivation to maintain and improve that environmental quality.

the economic and biophysical environment for all SESs, and strategies for interfacing an SES with the demands of a globalizing environment are needed (Young et al., 2006). Theories and models for the linkage of SESs across scales are also under development (Anderies, Folke, Walker, and Ostrom, 2013).

THE SOCIOECOLOGICAL SYSTEM HIERARCHY

The coupling of human social systems and ecological systems can be viewed from a hierarchical perspective. Each level in the nested hierarchy is an integrated system of social and biophysical components linked by causal relationships and feedbacks between the parts (Liu et al., 2007a, 2007b). Each level is also linked to other levels, usually operating at different spatial and temporal scales (Allen and Hoekstra, 1992). Optimally, there is a good match within an SES between the scale of the social system and the scale of the resource being managed (Cumming, Cumming, and Redman, 2006).

The Household

At the base of the SES hierarchy sits a household and an associated piece of land. "Piece of land" for the typical urbanite may reduce to a few potted plants, but in the case of a suburbanite in a developed country, this entity would be the combination of home, garden, and family. The energy flow into the household is in forms such as solar energy, electrical energy, and natural gas. The householder brings in food mechanically, and information electronically. There is export of wastes through the sewer pipes, the trash disposal service, and directly to the atmosphere. The ecosystem that is the garden depends on recurrent attention in the form of planting, weeding, and watering. The garden produces flowers, fruits, and vegetables, which provide physical and psychological sustenance. The garden has its own biogeochemical cycling of nutrients that is supplemented by synthetic fertilizers and compost.

A farmer and his or her farm, or a woodlot owner, is a step up in terms of scale. The farmer and co-workers may use heavy equipment, fossil fuels, fertilizers, pesticides, herbicides, and irrigation. Social interactions among co-workers may take various forms. Issues at this level of the resource management hierarchy revolve around understanding ecosystem processes sufficiently to manipulate them and extract surplus production without degrading the resources (especially soil properties). Understanding ecosystem metabolism could be by way of folk knowledge, as in slash and burn agriculture, or by way of formal scientific assessments of biogeochemical dynamics and community ecology (Chapin et al., 2010). Farmers encounter a very direct negative feedback loop if their management techniques reduce the productivity of the land.

The Landscape

The next level up from the householder and farm scales of management is the landscape scale; it may offer the clearest expression of the SES concept. A landscape is a mosaic of land cover and land use types. These polygons are management units, and they interact by transfers of mass (e.g., water), organisms (e.g., seeds), and disturbances (e.g., fire). A watershed or river basin is an especially appropriate domain for an SES because of its naturally defined borders. Landscape ecology is a relatively young field that has only recently begun to address the many design issues required to achieve sustainability at the landscape scale (Lindenmayer et al., 2008; Naveh, 2007; M. G. Turner, Gardner, and O'Neill, 2003).

Traditionally, rural landscapes in the United States were in some senses managed by associations of farmers or ranchers. The grange or the cattleman's association met in part to work out management of CPRs. More recently we have watershed councils (e.g., the Mary's River Watershed Council, covering the area in which I live; MRWC, 2017) in which stakeholders periodically meet, discuss issues, and initiate projects to promote sustainable management. Stakeholders include land owners, as well as representatives of public agencies such as the local Soil and Water Conservation District. River basins such as the Mekong River and Columbia River have become the foci of SESs that integrate across national borders.

Stakeholder involvement in the planning process is a key feature of an SES. Public land management agencies in the United States have made a strong commitment to stakeholder involvement. For example, the U.S. Forest Service develops and periodically updates management plans at the level of national forests (USDA, 2012). These plans cover multiple uses—including timber harvests, recreation, and biodiversity conservation—and hence can be contentious. The planning process usually involves face-to-face meetings of stakeholders, development of alternative scenarios, and attempts to reach consensus (USDA, 2016a).

Strong digital (geographic information systems) representation of the land base helps support the decision-making process. Landscape simulation models—with the capacity to periodically update land cover/land use classification based on factors such as urbanization and forest management—have also been implemented to facilitate scenario development (Bolte, Hulse, Gregory, and Smith, 2007; Rammer and Seidl, 2015).

Increasingly, urban landscapes are actively managed. Cities have become nodes in the global infrastructure that supports the production and consumption of human-made goods and services, including information. Resources such as energy, water, raw materials, crops, and livestock pour into cities. Manufactured goods and high-quality information come out, along with waste products in the form of low-quality energy and materials that enter the surrounding air, water, and soil. Some of the critical environmental management issues in urban landscapes are the number, size, and distribution of public parks; the degree of zonation; the establishment of an urban growth boundary; the emphasis on public versus private transportation; and issues with urban forestry. Management institutions include city councils and commissions, as well as the city agencies responsible for implementation of the agreed-upon plans.

Each SES covering a relatively small geographic domain is effectively an experiment in design and operation of such systems. Civil society networks (introduced in chapter 8) link across local SESs and help spread innovative ideas and successful practices. For example, several states in the United States have watershed coalitions that support community-led efforts to manage local water and land use issues (MWCC, 2017; NOWC, 2017).

The Bioregion

The bioregion is the next level up in the SES hierarchy. A bioregion has a characteristic climate, geomorphology, vegetation, and wildlife (McGinnis,

1998); generalizations can be made about the major adaptations of the plants and animals to the bioregional climate, as well as about how people support themselves and interact with the environment. The bioregion is a useful level of integration for the purposes of human management of natural resources because of its uniformity in climate, vegetation, and natural resources. As with landscape ecology, the field of regional ecology is newly developed. Progress in this field has been driven by challenges such as conservation of old-growth forests in the Pacific Northwest region of the United States (Thomas, Franklin, Gordon, and Johnson, 2006) and minimizing runoff of fertilizer nitrogen in the Midwest agricultural region (the Corn Belt) of the United States.

In my bioregion (Cascadia), the conservation of anadromous salmon has revealed the exceedingly wide scope of thinking necessary to address resource management issues at the bioregional scale. Salmon management requires regulating ocean and in-stream fishing, understanding climatic influences on fish populations, and regulating river flows to facilitate passage of spawning fish upriver and passage of smolts back downriver to the ocean. Regulation of logging in riparian areas is also needed because of the effects of the streamside trees on water temperature and effects of erosion from upland forests on gravel bars where spawning occurs. Anadromous fish effectively knit the entire bioregion together and their survival clearly requires compromises that may restrict the satisfaction of some human wants and needs.

The mechanisms of environmental governance at the level of bioregions and river basins are often problematic, a key issue being that political boundaries (be they county, state, province, or national) do not necessarily map well onto bioregional boundaries. For example, land ownership is often split between the public and private domains. Arriving at a management plan thus often requires agreements among governmental entities, each of which have heretofore had their own way. Assuming a reasonably democratic environment, governance structures can be built based on agreements within groups of stakeholders, i.e., all interested parties (public and private). Inevitably, however, decisions covering large spatial domains are hard fought and are rarely win–win. Thus, they may end up as weak or ineffective compromises.

Besides formal regulation of natural resources, the bioregion concept also evokes a bond between individuals and their environment. To create a bioregional identity and inspire participation in stakeholder forums (or financial support for nongovernmental organizations [NGOs] that represent them), the residents must come to identify with its particular geography. However, to the degree that bioregionalism develops today, it does so in the face of the great homogenizing force of modern consumerism. We can buy locally grown produce at the food co-op in town, or buy it cheaper from the large chain store down the road. The mega store is likely getting the produce from somewhere thousands of kilometers away. For entertainment, we can go to the local bar and see live music, or sit at home and watch a Hollywood movie on TV. For bioregionalism to become a significant trend, a conscious choice must be made to eat locally grown food, vacation locally, learn the local geology and botany, value the local culture, and value the local environment.

National Scale

The national scale as a level in the socioecological hierarchy is somewhat artificial from the perspective of biophysical processes because national boundaries are usually drawn more for geopolitical reasons than ecological reasons. But in the reality of how natural resources are managed, the national scale is critical, and will continue to be so in the coming decades. The progress of ecological modernization is evident at the national scale in the form of institutions, such as environmental protection agencies, as well as proliferation of environmentally oriented NGOs that lobby for environmental conservation in legislative arenas.

Enlightened management of natural resources (i.e., engagement of the social system with the physical/biological environment) requires a suite of science-based activities, and national governments have the financial resources and mandate to perform them. At the most basic level, natural resources must be inventoried and monitored. Because of the magnitude of this task, dedicated agencies and sustained research programs are needed. Ongoing monitoring efforts in the United States, for example, include projects that treat climate, atmospheric chemistry, stream flow, stream chemistry, land cover, land use, crop productivity, forest productivity, wildlife populations, and much more (see chapter 9). Development of environmental policy requires monitoring the status and detecting the trends in critical indicators related to each of these resources. Enforcement of infractions against environmental protection laws also depends on an effective monitoring system.

Besides monitoring capabilities, national agencies responsible for the long-term conservation of natural resources must have simulation-modeling programs that account for potential biophysical and socioeconomic changes affecting the resource. Preparing for climate change is a particularly relevant theme lately, but scenario building based on simulation modeling is desirable in relation to all threats to natural resources. Every five years, the U.S. Forest Service produces a fifty-year scenario of wood supply and demand for the U.S. forest sector that evaluates the effects of factors including economics, population size, land use change, and climate change (USDA, 2016b). Similar approaches are needed, but are rarely in place, in all countries for resources associated with the water, land, air, and ocean.

Even liberal democratic states have an inherent weakness with respect to management of the environment in that these states were initially designed primarily to facilitate "strategic bargaining or power trading among self-interested actors in the marketplace" (Eckersley, 2004, p. 115). The evolution of the state toward greater responsibility for the environment, especially shared natural resources, is thus an ongoing process. Environmental management at the national level is often delegated to various agencies or branches of government, which are to some degree isolated from the broad array of potential stakeholders with interest in those resources. A variety of innovative national-level approaches to environmental governance are under development that link national, private sector, and civil society actors (Galli and Fisher, 2016; Jänicke and Jörgens, 2009). This trend is consistent with the "strong" version of ecological modernization, which includes questioning the structure and function of societal institutions (Christoff, 1996).

Global Scale

The highest level in the socioecological hierarchy encompasses Earth as a whole. Conceptually, there is perhaps not a great fit here because often a key ingredient to the success of an SES is involvement of stakeholders with local knowledge. Also, natural resource management decisions are usually made at local to regional scales rather than the global scale. Nevertheless, environmental change issues are manifest at the global scale (e.g., climate change) and the SES model may provide some benefits.

The most obvious resources at the global scale are the CPRs of the atmosphere and the oceans. The global atmosphere is readily available for disposal of CO_2 associated with fossil fuel combustion, as well as other greenhouse gases, and industry-related toxic substances. Energy production benefits accrue to fossil fuel users, and the costs go to the global community in the form of climate change. The open oceans are not regulated as national territory, and hence can be exploited by overfishing and waste disposal. The user community for the atmosphere and oceans includes everyone on Earth. Biodiversity in the oceans and on land is another resource that could be considered global (e.g., conservation of animal species that migrate across continents).

Several obstacles have been identified when seeking to apply SES theory to management of global-scale CPRs (Challies, Newig, and Lenschow, 2014; Ostrom et al., 1999).

1. The large number of participants prevents the kind of face-to-face interactions that facilitate local-scale SESs.

2. The inclusion of participants from multiple cultural backgrounds makes communication, and reaching unanimous agreements, more difficult.

3. Individuals are less obviously dependent on the resource (e.g., the atmosphere) than in a small-scale, tightly coupled SES. Thus, willingness to sacrifice is limited.

4. No single CPR can be isolated; hence, management plans are necessarily complex and difficult to implement.

5. As the scale increases, the opportunities to experiment (i.e., adaptive management) decline.

A core conclusion about engaging the SES concept at the global scale is that a polycentric management is needed (Ostrom, 2010); i.e., global scale issues must be addressed at multiple levels in the SES hierarchy (Young et al., 2006). The nested nature of the hierarchy helps low-level SESs frame their contribution to addressing the issue. At high levels, SESs analyze environmental problems at their most comprehensive scale. The challenge, as noted earlier, is to develop social subsystems (institutions in some cases) at the right scale and with the right set of responsibilities to match bio-physical subsystems (Cumming et al., 2006).

The management of biodiversity is beginning to be addressed across a range of SESs. At the global scale, representatives of 152 nations recently met and signed an agreement to limit international trade in endangered species (CITES, 2017). Within nations, agencies like the U.S. Department of the Interior (DOI) have mandates to identify endangered species and devise conservation plans. Another level below the DOI is the network of landscape conservation cooperatives (LCC, 2016) supported by multiple governmental agencies. At even lower levels there might be, for example, a biodiversity task force at the city level. From the polycentric management perspective, a World Environment Organization (see chapter 8) could help coordinate across levels.

The issue of reducing greenhouse gas emissions is also representative here. Despite the lack of immediate payoff in terms of altering climate change, and the obvious massive free rider implications, actions to reduce emissions are now being taken by many individuals, cities, states, regions, and nations (Ostrom, 2010). The motivations trace to a variety of potential benefits, such as (1) improved health from use of alternatives to automobile transportation (e.g., bicycles), (2) improved health from reductions in air pollution, and (3) supporting industries to take advantage of trends in renewable energy generation. The polycentric approach facilitates experiments in technology and policy.

SOCIOECOLOGICAL SYSTEM MANAGEMENT OF MULTIPLE ECOSYSTEM SERVICES

The complexity of organizing and operating an SES that is focused on one resource (e.g., water) is challenging. However, in practice, society depends on a wide range of ecosystem services (MEA, 2003), and there are often tradeoffs among them when management interventions are planned (Bennett, Peterson, and Gordon, 2009); e.g., conservation of biodiversity may mean setting aside land that could potentially be used for agriculture, grazing, or forestry. Policy analysts have identified the issue of "problem shifting," which occurs when improvements made in management of one resource cause degradation of another (Nilsson and Persson, 2012). In some cases, ecosystem services are synergistic and can be effectively "bundled" for management purposes (Raudsepp-Hearne, Peterson, and Bennett, 2010a). But, in any case, working out the economic aspects of such trade-offs is challenging.

On the social system side of an SES, treating more resources means involving a greater diversity of interests and institutions. Research on the social subsystem side of SESs is providing insights into factors such as the level of engagement of the stakeholders (consultative versus functional), the level of consensus required, the nature of the power relations among participants, and the timing of stakeholder involvement (Reed, 2008).

Trade-offs in ecosystem services at the global scale are especially pertinent in the context of climate change mitigation and adaptation (Scholes, 2016). The IPCC stabilization scenario (described in chapter 6) depends on extensive biomass energy, combined with carbon capture and sequestration, to draw down the atmospheric CO_2 concentration. However, the area of land required begins to impinge on the land available for provision of food and fiber (Bierbaum and Matson, 2013). As yet, there is limited capacity to quantitatively evaluate these trade-offs (e.g., Naidoo et al., 2008) and deliberate about them as a global SES. Integrated assessment models (see chapter 6) are an operational tool to link across resources of this kind (van Vuuren et al., 2015).

The U.S. National Science Foundation has recently instituted a program focused on "Innovations at the Nexus of Food, Energy and Water Systems" (INFEWS, 2017). The program supports inter- and transdisciplinary teams that address complex real-world natural resource management issues. New modeling frameworks are emerging to handle "the integrative, adaptive, hierarchical, complex and multiscale nature of interdependencies" among the biophysical and social components of SESs (Giampietro, Aspinall, Ramos-Martin, and Bukkens, 2014, p. 1).

Progress at managing multiple ecosystem services is most advanced at the landscape to bioregion scale, and a key tool for integration across multiple ecosystem services is landscape simulation modeling. Spatially explicit representation of resources across the landscape, and of the associated process by which the resources are maintained and changed, provides a framework for assessment of current status, historical trends, and future possibilities for multiple resources. Land cover is a principal factor in provision of many ecosystem services (Nelson et al., 2009), and a state and transition model (Rounsevell, Robinson, and Murray-Rust, 2012) can periodically update land cover classification based on modeled transition probabilities (e.g., a prescribed rate of logging over a forest landscape).

As scientific understanding and computing power increase, the range of ecosystem services covered within a harmonized landscape modeling infrastructure is expanding. In a recently developed SES at the river basin scale, the modeling framework accounted for the effects of land use change, forest management, and climate change on water scarcity (WW2100, 2017). Stakeholders included local public utility water managers, reservoir operators, and regional forest managers. Deliberations covered, for example, the effect of climate on wildfire as well as the effect of wildfire suppression on streamflow (D. P. Turner, Conklin, Bolte, 2015; D. P. Turner et al., 2016).

ESSENTIAL POINTS

The era of exploiting natural resources without concern for sustainability is rapidly ending. In its place must come a hierarchy of SESs that extend from the level of local ecosystems, through to landscapes, bioregions, nations, and the planet. Management at any one level must be cognizant of and integrated with levels above and below. This polycentric management configuration allows for experimentation at relatively low levels, consistent with the kind of adaptive management approach that will be needed in a rapidly changing environment. Given the simultaneous encroachment on multiple planetary boundaries, it will be necessary to account for tradeoffs among ecosystem services in SES-based management efforts.

IMPLICATIONS

1. Effective natural resource management often involves the coupling of a social system, which includes multiple stakeholders, with an ecological system.

2. The appropriate geographic scale for managing a natural resource varies widely depending on the resource.

3. SESs require the capacity to monitor the relevant resources, as well as decision support tools such as landscape simulation models.

4. Adaptive management may be necessary for the operation of complex SESs, but it is best employed at relatively low levels of a socioecological hierarchy to allow for failures.

FURTHER READING

Chapin, F. S., Carpenter, S. R., Kofinas, G. P., Folke, C., Abel, N., Clark, W. C., . . . Swanson, F. J. (2010). Ecosystem stewardship: Sustainability strategies for a rapidly changing planet. *Trends in Ecology and Evolution, 25,* 241–249.

Folke, C., Jansson, A., Rockström, J., Olsson, P., Carpenter, S. R., Chapin, F. S., . . . Westley, F. (2011). Reconnecting to the biosphere. *Ambio, 40,* 719–738.

11

KEY CONCEPTS FOR A NEW PLANETARY PARADIGM

We have seen that the impacts of global environmental change are beginning to be experienced widely, and that the implications for business-as-usual expansion of the technosphere are becoming ever clearer. The familiar planetary paradigm of humanity as insignificant relative to the background operation of the Earth system is obsolete, and a new planetary paradigm that places humanity as a core component in the structure and function of the Earth system is needed. By way of cultural evolution, this new planetary paradigm is currently being conceived and propagated. Here we will revisit and integrate several key concepts introduced earlier that are relevant to this project.

THE ANTHROPOCENE NARRATIVE

Humans are storytelling animals. Our brains are wired to assimilate information in terms of temporal sequences of significant events. We are likewise cultural animals. Within a society, we share images, words, rituals, and stories. Indigenous societies often have myths about their origin and history. Religious mythologies remain prevalent in contemporary societies.

Earth system science has revealed the need to self-organize a global society that will address emerging global environmental change issues. A shared narrative about the relationship of humanity to the biosphere, and more broadly to the Earth system, is certainly desirable in that context. Hence, the Anthropocene narrative.

Societal narratives tend to have normative implications. For example, the "grand narrative" of the progress of western civilization implied a certain virtue about features such as colonialism and an economic system based on the self-regulated market. Likewise, a societal narrative may ignore significant divisions or inequalities within that society. These aspects of societal narratives are often not questioned. But in keeping with the call by environmental sociologists for a second, reflexive, modernization, let's make the Anthropocene narrative a reflexive narrative. It is a story of humanity's relationship to the global environment, but the ending is uncertain and questions are posed about what humanity is now doing to the biosphere, as well as what is needed to achieve a sustainable Earth system.

Recall the stages of the Anthropocene narrative introduced in chapter 1 (table 1.1).

The Prehuman Gaian Biosphere

The Gaian biosphere, which self-organized relatively quickly after the coalescence of the geosphere, is fueled mostly by solar energy. The biosphere drives the global biogeochemical cycles of carbon, nitrogen, and other elements essential to life, and plays a significant role in regulating Earth's climate, as well as the chemistry of the atmosphere and oceans. The biosphere augments a key geochemical feedback mechanism in the Earth system (the rock weathering thermostat) that has helped keep the planet's climate in the habitable range for four billion years. By way of collisions with comets or asteroids, or because of its own internal dynamics, the Earth system occasionally reverts to conditions that are harsh for many life forms (i.e., extinction events). Nevertheless, the biosphere has always recovered—by way of biological evolution—and recently produced a mammalian primate species that was qualitatively different from any of its predecessors.

The Great Separation

Nervous systems in animals have obvious adaptive significance in terms of sensing the environment and coordinating behavior. The brain of a human being appears to be a rather hypertrophied organ of the nervous system, but has evolved in support of a capacity for language and self-awareness. These capabilities are quite distinctive among animal species and set the stage for the human conquest of the planet. As the most recent ice age receded about 10,000 years ago, *Homo sapiens* emerged from Africa and dispersed widely. An unusually stable Holocene climate favored the discovery and expansion of agriculture. With agriculture, and gradual elaboration of toolmaking, humanity ceased waiting for Nature to provide it sustenance. Rather, Nature became an object to be managed. This change is, of course, also captured in the religious myth of the Garden of Eden.

Building the Technosphere

The next phase in the Anthropocene narrative is characterized by the ascent of the scientific worldview, the spread of representative government, and the establishment of the market system. Human population rose to the range of billions and the technosphere began to cloak Earth. The Industrial Revolution vastly increased the rate of energy flow and materials cycling by the human enterprise. The telecommunications and transportation infrastructures expanded and humanity began to get a sense of itself as a global entity. Evidence that humans could locally overexploit natural resources (e.g., the runs of anadromous salmon in the Pacific Northwest of the United States) began to accumulate.

The Great Acceleration

Between World War II and the present, the global population grew to more than 7.5 billion. Scientific advances in the medical field reduced human mortality rates, and technical advances in agriculture, forestry, and fish harvesting largely kept pace with provision of food and fiber. The density of the technosphere increased rapidly. At the same time, we began to see evidence of technosphere impacts on the environment at the global scale—notably, changes in atmospheric chemistry and losses in global biodiversity.

The Great Transition

This phase is just beginning. Its dominant signal will be the bending of the exponentially rising curves for Earth system and socioeconomic indicators (see figures 4.2 and 4.3). The Great Transition is envisioned to be accomplished within the framework of a high-technology infrastructure and robust global economy. To accomplish this multigenerational task, humanity must begin to function as a global-scale collective, capable of self-regulating.

Equilibrium

Human-induced global environmental change will continue for the foreseeable future. The assumption for an Equilibrium phase is that humanity will gain a good understanding of the Earth system—including the climate subsystem and the global biogeochemical cycles—and develop sufficiently advanced technology to begin managing the technosphere and biosphere for the purposes of shaping the global biophysical environment. Humanity itself is, of course, part of the Earth system, meaning it must gain sufficient understanding of the social sciences to produce successive generations of global citizens who value environmental quality and will cooperate to manage and maintain it.

The Anthropocene narrative is broadly consistent with scientific observations and theories, which gives it a chance for wide acceptance. It also provides a solid rationale for building a global community of all human beings, as we are all faced with the challenge of living together on a crowded planet. The assimilation of the Anthropocene narrative, or something like it, by a sizable proportion of the global population would seem to be a requirement for the eventual materialization of its later stages.

THE ICARUS SCENARIO

The study of Earth's environmental history in the paleorecord reveals episodes in which an elevated level of greenhouse gas emissions from the geosphere led to high atmospheric concentrations of those gases, and radical changes in the physical climate, the biosphere, and the global biogeochemical cycles. Slowing of ocean circulation and stratification of the ocean appear to be significant contributors to the associated extinction events. The eerily similar anthropogenic pumping of greenhouse gases into the atmosphere could, if unabated, push the climate system through multiple tipping points and, over hundreds to thousands of years, to a repeat of the earlier global environmental convulsions.

The point of evoking the Icarus scenario is to emphasize that we know enough about the Earth system to imagine it happening, but we don't know enough about the Earth system and our own self-destructive tendencies to rule it out. Thus, there is a strong basis for altering the current trajectory toward a massive disruption of the Earth system and widespread human misery.

THE NOÖSPHERE SCENARIO

In chapter 1, we settled on a definition for noösphere as a planet on which a biosphere has given rise to a sentient species that has managed to build a technosphere and establish a sustainable, high-technology, global civilization (i.e., playing out the Anthropocene narrative). There are obviously many significant challenges along the way to building a noösphere, but given the order-friendly universe we live in—and how far we have come—humanity can reasonably aspire to accomplishing it. Doing so would require stronger institutions for global governance, especially a new intergovernmental institution for global environmental governance—a World Environment Organization.

ECOLOGICAL MODERNIZATION

This grand abstraction refers to the process by which environmental considerations rise to the level of factors such as economics, security, and social justice in all societal deliberations. Ecological modernization (EM) does not involve radically altering current sociopolitical and socioeconomic systems, but rather developing appropriate technologies and conservation-oriented natural resource management schemes. EM is a decidedly bottom-up approach, including vital civil society and corporate responsibility movements. However, it also relies on robust governmental regulation when needed. EM is energized by consumers when they select products that are certified "green," and voters when they support politicians who follow an environmentally friendly agenda. EM results in policies that (1) end externalization of costs by business interests, (2) foster eco-friendly technologies (e.g., renewable energy sources), and (3) steer land-use toward enhancement of biodiversity. The geographic scale at which EM must be actualized extends from local to global.

LEXICON OF THE SPHERES

ANTHROPOSPHERE: "Human societies, cultures, knowledge, economies and built environments" (Raupach and Canadell, 2010, p. 210).

ATMOSPHERE: A layer of gases surrounding Earth and retained by the planet's gravity.

BIOSPHERE: The sum total of all living things on Earth. It was originally defined by Suess as the place on Earth's surface where life dwells. However, in the context of biogeochemical cycling (according to Vernadsky, 1945), it is more appropriate to identify it as an evolving life system that operates as a geological force transforming the planet's surface and maintaining the global biogeochemical cycles.

CRYOSPHERE: The portions of the Earth system where water is in a solid form, including snow, ice, glaciers, sea ice, and permafrost.

CYBERSPHERE: All information and contacts available over the Internet.

ECOSPHERE: Encompassing both the biological and physical components of the planet, the ecosphere initially was the purely physical geosphere, atmosphere, and hydrosphere. Over the last four billion years it has come to include the biosphere and the technosphere (Gillard, 1969).

GAIA: The living Earth system, including geosphere, atmosphere, hydrosphere, and biosphere.

GEOSPHERE: Broadly defined as the solid part of Earth. More formally it can be differentiated into the lithosphere (surface, crust, and mantle) and the core.

HYDROSPHERE: The water on Earth in all forms: the ocean (which is the bulk of the hydrosphere); other surface waters, including inland seas, lakes, and rivers; rain; underground water; ice (as in glaciers and snow); and atmospheric water vapor (as in clouds).

IONOSPHERE: The uppermost part of the atmosphere, it can be distinguished from lower layers because atoms and molecules are ionized by solar radiation.

LITHOSPHERE: The outermost layer of rock on the planet. The lithosphere is divided into tectonic plates that move independently of each other.

MAGNETOSPHERE: The region around Earth in which phenomena are dominated or organized by its magnetic field.

NOÖSPHERE: The Earth system under the influence of a conscious (self-aware) humanity. There are alternative interpretations of the concept from Teilhard de Chardin and Vernadsky (D. P. Turner, 2005).

PEDOSPHERE: The layer of soil on Earth's surface.

PHYSIOSPHERE: All material components of the Earth system.

SEMIOSPHERE: The sphere of signs and symbols. The concept is developed in a collection of Lotman's writings, published in English under the title *Universe of the Mind: A Semiotic Theory of Culture* (Lotman, 1990).

SOCIOSPHERE: All human beings on the planet and all their interrelationships.

STRATOSPHERE: The second layer of Earth's atmosphere, just above the troposphere and below the mesosphere. Important in relation to the biosphere because ozone in the stratosphere absorbs most of the damaging ultraviolet radiation in unfiltered sunlight.

TECHNOBIOSPHERE: The result of integration of the human-generated technosphere with Earth's biosphere (D. P. Turner, 2011). The technobiosphere now influences global atmospheric chemistry and climate.

TECHNOSPHERE: The sum of all human artifacts, including buildings, roads, machines, and electronic devices. It is maintained by a flow of material and energy from the geosphere, biosphere, and sun (Haff, 2014).

TROPOSPHERE: The lowest layer of the atmosphere (heights to 20 kilometers). It contains approximately 75 percent of the atmosphere's mass and is involved in active exchange of mass and energy with the planet's surface.

REFERENCES

Abatzoglou, J. T., and Brown, T. J. (2011). A comparison of statistical downscaling methods suited for wildfire applications. *International Journal of Climatology, 32*, 772–780.

Abell, R., Lehner, B., Thieme, M., and Linke, S. (2016). Looking beyond the fenceline: Assessing protection gaps for the world's rivers. *Conservation Letters*. doi: 10.1111/conl.12312

Aber, J., McDowell, W., Nadelhoffer, K., Magill, A., Berntson, G., Kamakea, M., . . . Fernandez, I. (1998). Nitrogen saturation in temperate forest ecosystems— Hypotheses revisited. *Bioscience, 48*, 921–934.

Abruzzi, W. S. (1995). The social and ecological consequences of early cattle ranching in the Little Colorado River Basin. *Human Ecology, 23*, 75–98.

Acheson, J. M. (2003). *Capturing the commons: Devising institutions to manage the Maine lobster industry*, Lebanon, NH: University Press of New England.

Ackerman, F., DeCanio, S. J., Howarth, R. B., and Sheeran, K. (2009). Limitations of integrated assessment models of climate change. *Climatic Change, 95*, 297–315.

Adams, W. M., Brockington, D., Dyson, J., and Vira, B. (2003). Managing tragedies: Understanding conflict over common pool resources. *Science, 302*, 1915–1916.

Aklin, M., and Urpelainen, J. (2014). The global spread of environmental ministries: Domestic-international interactions. *International Studies Quarterly, 58*, 764–780.

Alcamo, J., Kok, K., Busc, G., Eickhout, B., Priess, J. A., Rounsevell, M., . . . Heisterman, M. (2006). Searching for the future of land: scenarios from the local to global scale. In E. Lambin and H. Geist (Eds.), *Land use and land cover change: Local processes, global impacts* (pp. 137–155). Berlin: Springer Verlag.

Alcott, B. (2005). Jevons' paradox. *Ecological Economics, 54*, 9–21.

Allen, C. D., Macalady, A. K., Chenchouni, H., Bachelet, D., McDowell, N., Vennetier, M., . . . Cobb, N. (2010). A global overview of drought and heat-induced tree mortality reveals emerging climate change risks for forests. *Forest Ecology and Management, 259*, 660–684.

Allen, T. F., and Hoekstra, T. (1992). *Toward a unified ecology*. New York: Columbia University Press.

Amos, H. M., Jacob, D. J., Streets, D. G., and Sunderland, E.M. (2013). Legacy impacts of all-time anthropogenic emissions on the global mercury cycle. *Global Biogeochemical Cycles, 27,* 410–421.

Anderegg, W. R. L., Ballantyne, A. P., Smith, W. K., Majkut, J., Rabin, S., Beaulieu, C., . . . Pacala, M. (2015). Tropical nighttime warming as a dominant driver of variability in the terrestrial carbon sink. *Proceedings of the National Academy of Sciences U.S.A., 112,* 15591–15596.

Anderies, J. M., Folke, C., Walker, B., and Ostrom, E. (2013). Aligning key concepts for global change policy: Robustness, resilience, and sustainability. *Ecology and Society, 18.* doi: 10.5751/es-05178–180208

Anderies, J. M., Walker, B. H., and Kinzig, A. P. (2006). Fifteen weddings and a funeral: Case studies and resilience-based management. *Ecology and Society, 11,* 21.

Angelstam, P., Andersson, K., Annerstedt, M., Axelsson, R., Elbakidze, M., Garrido, P., . . . Stjernquist, I. (2013). Solving problems in social-ecological systems: Definition, practice and barriers of transdisciplinary research. *Ambio, 42,* 254–265.

Archer, D. (2005). Fate of fossil fuel CO_2 in geologic time. *Journal of Geophysical Research-Oceans, 110.* doi: 10.1029/2004jc002625

Archer, D. (2010). *The global carbon cycle.* Princeton, NJ: Princeton University Press.

Archer, D., Buffett, B., and Brovkin, V. (2009). Ocean methane hydrates as a slow tipping point in the global carbon cycle. *Proceedings of the National Academy of Sciences U.S.A., 106,* 20596–20601.

ARGO. (2017). Argo: Part of the integrated global observation strategy. Retrieved from http://www.argo.ucsd.edu/

Armstrong, A. M., Wang, B. X. H., and Tang, C. C. (2015). Accelerating low-carbon development in Portland, Oregon and Kunming, Yunnan. *Journal of Renewable and Sustainable Energy, 7.* doi: 10.1063/1.4928418

Arnell, N. W., and Gosling, S. N. (2016). The impacts of climate change on river flood risk at the global scale. *Climatic Change, 134,* 387–401.

Asner, G. P., Elmore, A. J., Olander, L. P., Martin, R. E., and Harris, A. T. (2004). Grazing systems, ecosystem responses, and global change. *Annual Review of Environment and Resources, 29,* 261–299.

Asner, G. P., and Martin, R. E. (2009). Airborne spectranomics: Mapping canopy chemical and taxonomic diversity in tropical forests. *Frontiers in Ecology and the Environment, 7,* 269–276.

Atkinson, A., Siegel, V., Pakhomov, E., and Rothery, P. (2004). Long-term decline in krill stock and increase in salps within the Southern Ocean. *Nature, 432,* 100–103.

Azam, M., and Khan, A. Q. (2016). Testing the Environmental Kuznets Curve hypothesis: A comparative empirical study for low, lower middle, upper middle and high income countries. *Renewable and Sustainable Energy Reviews, 63,* 556–567.

Bachelet, D., and Turner, D. P. (Eds.). (2015). *Global vegetation dynamics.* Hoboken, NJ: Wiley.

Bak, P. (1996). *How nature works.* Göttingen, Sweden: Copernicus.

Baker, A. C., Glynn, P. W., and Riegl, B. (2008). Climate change and coral reef bleaching: An ecological assessment of long-term impacts, recovery trends and future outlook. *Estuary Coastal Shelf Science, 80,* 435–471.

Baldocchi, D. D. (2003). Assessing the eddy covariance technique for evaluating carbon dioxide exchange rates of ecosystems: past, present and future. *Global Change Biology, 9,* 479–492.

Ballantyne, A. P., Alden, C. B., Miller, J. B., Tans, P. P., and White, J. W. C. (2012). Increase in observed net carbon dioxide uptake by land and oceans during the past 50 years. *Nature, 488,* 70–73.

Barnett, T. P., Pierce, D. W., Hidalgo, H. G., Bonfils, C., Santer, B. D., Das, T., . . . Dettinger, M. D. (2008). Human-induced changes in the hydrology of the western United States. *Science, 319,* 1080–1083.

Barnosky, A. D., Matzke, N., Tomiya, S., Wogan, G. O. U., Swartz, B., Quental, T. B., . . . Ferrer, E. A. (2011). Has the Earth's sixth mass extinction already arrived? *Nature, 471,* 51–57.

Barnosky, A. D., Brown, J. H., Daily, G. C., Dirzo, R., Ehrlich, A. H., Ehrlich, P. R., . . . Wake, M. H. (2014). Introducing the scientific consensus on maintaining humanity's life support systems in the 21st century: Information for policy makers. *Anthropocene Review, 1,* 78–109.

Barrow, J. D. (2002). *The constants of nature: The numbers that encode the deepest secrets of the universe.* New York: Pantheon.

Bartoli, G., Sarnthein, M., Weinelt, M., Erlenkeuser, H., Garbe-Schonberg, D., and Lea, D. W. (2005). Final closure of Panama and the onset of northern hemisphere glaciation. *Earth and Planetary Science Letters, 237,* 33–44.

Bateson, G. (1979). *Mind and nature.* Hialeah, FL: Dutton Adult.

Beare, D. J., Burns, F., Greig, A., Jones, E. G., Peach, K., Kienzle, M., . . . Reid, D. G. (2004). Long-term increases in prevalence of North Sea fishes having southern biogeographic affinities. *Marine Ecology Progress Series, 284,* 269–278.

Beaugrand, G., Reid, P. C., Ibanez, F., Lindley, J. A., and Edwards, M. (2002). Reorganization of North Atlantic marine copepod biodiversity and climate. *Science, 296,* 1692–1694.

Beck, U. (1992). *Risk society: Towards a new modernity.* Thousand Oaks, CA: Sage.

Beck, U. (1999). *Global risk society.* Cambridge, MA: Polity Press.

Beck, U., Gidden, A., and Lash, S. (1994). *Reflexive modernization. Politics, tradition and aesthetics in the modern social order.* Redwood City, CA: Stanford University Press.

Beck, U., and Grande, E. (2010). Varieties of second modernity: The cosmopolitan turn in social and political theory and research. *British Journal of Sociology, 61,* 409–443.

Beerling, D. J., Hewitt, C. N., Pyle, J. A., and Raven, J. A. (2007). Critical issues in trace gas biogeochemistry and global change. *Philosophical Transactions of the Royal Society A–Mathematical Physical and Engineering Sciences, 365,* 1629–1642.

Beerling, D. J., and Royer, D. L. (2011). Convergent Cenozoic CO_2 history. *Nature Geoscience*, *4*, 418–420.

Beerling, D. J., Taylor, L. L., Bradshaw, C. D. C., Lunt, D. J., Valdes, P. J., Banwart, S. A., ... Leake, J. R. (2012). Ecosystem CO_2 starvation and terrestrial silicate weathering: Mechanisms and global-scale quantification during the late Miocene. *Journal of Ecology*, *100*, 31–41.

Beerling, D. J., and Woodward, F. I. (1997). Changes in land plant function over the Phanerozoic: Reconstructions based on the fossil record. *Botanical Journal Linnean Society*, *124*, 137–153.

Behrenfeld, M. J., O'Malley, R. T., Boss, E. S., Westberry, T. K., Graff, J. R., Halsey, K. H., ... Brown, M. B. (2016). Revaluating ocean warming impacts on global phytoplankton. *Nature Climate Change*, *6*, 323–330.

Bennett, E. M., Peterson, G. D., and Gordon, L. J. (2009). Understanding relationships among multiple ecosystem services. *Ecology Letters*, *12*, 1394–1404.

Bentz, B. J., Regniere, J., Fettig, C. J., Hansen, E. M., Hayes, J. L., Hicke, J. A., ... Seybold, S. J. (2010). Climate change and bark beetles of the western United States and Canada: Direct and indirect effects. *Bioscience*, *60*, 602–613.

Berger, A., Loutre, M. F., and Crucifix, M. (2003). The Earth's climate in the next hundred thousand years (100 kyr). *Surveys in Geophysics*, *24*, 117–138.

Beringer, J., Chapin, F. S., Thompson, C. C., and McGuire, A. D. (2005). Surface energy exchanges along a tundra-forest transition and feedbacks to climate. *Agricultural and Forest Meteorology*, *131*, 143–161.

Berkes, F., Colding, J., and Folke, C. (Eds.). (2003). *Navigating socio-ecological systems: Building resilience for complexity and change*. Cambridge: Cambridge University Press.

Bertram, C., Luderer, G., Pietzcker, R. C., Schmid, E., Kriegler, E., and Edenhofer, O. (2015). Complementing carbon prices with technology policies to keep climate targets within reach. *Nature Climate Chang*, *5*, 235–239.

Bierbaum, R. M., and Matson, P. A. (2013). Energy in the context of sustainability. *Daedalus*, *142*, 146–161.

Biermann, F. (2007). Earth system governance as a crosscutting theme of global change research. *Global Environmental Change: Human and Policy Dimensions*, *17*, 326–337.

Biermann, F. (2012). Planetary boundaries and earth system governance: Exploring the links. *Ecological Economics*, *81*, 4–9.

Biermann, F. (2014). *Earth system governance: World politics in the Anthropocene*. Cambridge, MA: MIT Press.

Biermann, F., and Bauer, S. (Eds.). (2005). *A world environment organization*. London: Routledge.

Bloomberg. (2015). In coal setback, rich nations agree to end export credits. Retrieved from http://www.bloomberg.com/news/articles/2015-11-18/in-latest-blow-to-coal-rich-nations-agree-to-end-export-credits

Boden, T. A., Marland, G., and Andres, R. J. (2015). *National CO2 emissions from fossil-fuel burning, cement manufacture, and gas flaring: 1751–2011.* Carbon Dioxide Information Analysis Center, Oak Ridge National Laboratory, U.S. Department of Energy. doi:10.3334/CDIAC/00001_V2015

Bolte, J. P., Hulse, D. W., Gregory, S. V., and Smith, C. (2007). Modeling biocomplexity—actors, landscapes and alternative futures. *Environmental Modeling Software, 22,* 570–579.

Boren, Z. D. (2014). Active mobile phones outnumber humans for the first time. *International Business Times,* October 7. Retrieved from http://www.ibtimes.co.uk/there-are-more-gadgets-there-are-people-world-1468947

Boulding, K. (1964). *The meaning of the 20th century: The Great Transition.* New York: Harper Colophon.

Boulding, K. (1965). Earth as a spaceship. Lecture at Washington State University. https://bertaux.files.wordpress.com/2014/08/boulding-earth-as-spaceship-1965.pdf

Boutron, C. F. (1995). Historical reconstruction of the Earth's past atmospheric environment from Greenland and Antarctic snow and ice cores. *Environmental Review, 3,* 1–28.

Boyce, D. G., Lewis, M. R., and Worm, B. (2010). Global phytoplankton decline over the past century. *Nature, 469,* 591–596.

Boyce, J. K. (2004). Green and brown? Globalization and the environment. *Oxford Review of Economic Policy, 20,* 105–128.

Breshears, D. D., Cobb, N. S., Rich, P. M., Price, K. P., Allen, C. D., Balice, R. G., . . . Meyer, C. W. (2005). Regional vegetation die-off in response to global-change-type drought. *Proceedings of the National Academy of Sciences U.S.A., 102,* 15144–15148.

Broecker, W. S. (1987). The great ocean conveyor. *Natural History, 97,* 74–82.

Broecker, W. S. (2008). CO$_2$ capture and storage: Possibilities and perspectives. *Elements, 4,* 295–297.

Bronowski, J. (1974). *The ascent of man.* New York: Little, Brown.

Bruno, J. F., and Selig, E. R. (2007). Regional decline of coral cover in the Indo-Pacific: Timing, extent, and subregional comparisons. *Plos One.* doi: 10.1371/journal.pone.0000711

Budyko, M. I. (1984). *The evolution of the biosphere.* Dordrecht, Germany: D. Reidel.

Buermann, W., Beaulieu, C., Parida, B., Medvigy, D., Collatz, G. J., Sheffield, J., and Sarmiento, J. L. (2016). Climate-driven shifts in continental net primary production implicated as a driver of a recent abrupt increase in the land carbon sink. *Biogeosciences, 13,* 1597–1607.

Cai, W. J., Borlace, S., Lengaigne, M., van Rensch, P., Collins, M., Vecchi, G., . . . Jin, F. F. (2014). Increasing frequency of extreme El Nino events due to greenhouse warming. *Nature Climate Change, 4,* 111–116.

Cai, W. J., Santoso, A., Wang, G. J., Yeh, S. W., An, S . I., Cobb, K. M., . . . Wu, L. (2015). ENSO and greenhouse warming. *Nature Climate Change, 5,* 849–859.

Cai, Y. Y., Lenton, T. M., and Lontzek, T. S. (2016). Risk of multiple interacting tipping points should encourage rapid CO_2 emission reduction. *Nature Climate Change, 6,* 520–525.

Caldeira, K., and Wickett, M. E. (2005). Ocean model predictions of chemistry changes from carbon dioxide emissions to the atmosphere and ocean. *Journal of Geophysical Research-Oceans, 110.* doi: 10.1029/2004jc002671

Calvin, K., Edmonds, J., Bond-Lamberty, B., Clarke, L., Kim, S. H., Kyle, P., . . . Wise, M. (2009). Limiting, climate change to 450 ppm CO_2 equivalent in the 21st century. *Energy Economics, 31,* S107–S120.

CarbonTracker. (2017). Earth System Research Laboratory Global Monitoring Division, CarbonTracker CT2016. Retrieved from http://www.esrl.noaa.gov/gmd/ccgg/carbontracker/

Carroll, A. B. (1999). Corporate social responsibility. *Business and Society, 38,* 268–295.

Carson, R. (1962). *Silent spring.* Boston: Houghton Mifflin.

Casey, G., and Galor, O. (2017). Is faster economic growth compatible with reductions in carbon emissions? The role of diminished population growth. *Environmental Research Letters, 12,* 014003.

Castree, N. (2016). Broaden research on the human dimensions of climate change. *Nature Climate Change, 6,* 731.

CBD. (2017). Convention on Biological Diversity. Retrieved from https://www.cbd.int/

Ceballos, G., Ehrlich, P. R., Barnosky, A. D., Garcia, A., Pringle, R. M., and Palmer, T. M. (2015). Accelerated modern human-induced species losses: Entering the sixth mass extinction. *Science Advances 2015.* doi: 10.1126/sciadv.1400253

Ceccato, P., and Dinku, T. (2016). *Introduction to remote sensing for monitoring rainfall, temperature, vegetation and water bodies.* (IRI Technical Report 10–04). New York: Columbia University, International Research Institute for Climate and Society Earth Institute.

CEOS. (2017). Committee on Earth Observation Satellites. Retrieved from http://ceos.org/

CESM. (2017). Community Earth System Model. Retrieved from http://www2.cesm.ucar.edu/

Cess, R. D., Zhang, M. H., Ingram, W. J., Potter, G. L., Alskseev, V., Barker, H. W., . . . Wetherald, R. T. (1996). Cloud feedback in atmospheric general circulation models: An update. *Journal of Geophysical Research-Atmospheres, 101,* 12791–12794.

Chaisson, E. J. (2002). *Cosmic evolution: The rise of complexity in nature.* Boston: Harvard University Press.

Challies, E., Newig, J., and Lenschow, A. (2014). What role for social-ecological systems research in governing global teleconnections? *Global Environmental Change: Human and Policy Dimensions, 27,* 32–40.

Chan, K. M. A., Balvanera, P., Benessaiah, K., Chapman, M., Diaz, S., Gomez-Baggethun, E., . . . Turner, N. (2016). Opinion: Why protect nature? Rethinking values and the environment. *Proceedings of the National Academy of Sciences U.S.A., 113*, 1462–1465.

Chang, E. K. M., Guo, Y. J., and Xia, X. M. (2012). CMIP5 multimodel ensemble projection of storm track change under global warming. *Journal of Geophysical Research–Atmospheres, 117*. doi: 10.1029/2012jd018578

Chapin, F. S., Carpenter, S. R., Kofinas, G. P., Folke, C., Abel, N., Clark, W. C., . . . Swanson, F. J. (2010). Ecosystem stewardship: Sustainability strategies for a rapidly changing planet. *Trends in Ecology and Evolution, 25*, 241–249.

Charlson, R. J., Lovelock, J. E., Andreae, M. O., and Warren, S. J. (1987). Ocean phytoplankton, atmospheric sulfur, cloud albedo, and climate. *Nature, 326*, 655–661.

Chase-Dunn, C., and Lerro, B. (2013). *Social change: Globalization from the Stone Age to the present.* Boulder, CO: Paradigm.

Chatfield, R. B. (1991). Ephemeral biogenic emissions and the Earth's radiative and oxidative environment. In S. H. Schneider and P. J. Boston (Eds.), *Scientists on Gaia* (pp. 296–308). Cambridge, MA: MIT Press.

Chirot, D., and Hall, T. D. (1982). World-system theory. *Annual Review of Sociology, 8*, 81–106.

Christian, C., Ainley, D., Bailey, M., Dayton, P., Hocevar, J., LeVine, M., . . . Jacquet, J. (2013). A review of formal objections to Marine Stewardship Council fisheries certifications. *Biological Conservation, 161*, 10–17.

Christian, D., Brown, C., and Benjamin, C. (2013). *Big history: Between nothing and everything.* New York: McGraw-Hill.

Christoff, P. (1996). Ecological modernization, ecological modernities. *Environmental Politics, 5*, 476–500.

Chung, E. S., Soden, B., Sohn, B. J., and Shi, L. (2014). Upper-tropospheric moistening in response to anthropogenic warming. *Proceedings of the National Academy of Sciences U.S.A., 111*, 11636–11641.

Ciais, P., Dolman, A. J., Bombelli, A., Duren, R., Peregon, A., Rayner, P. J., . . . Zehner, C. (2014). Current systematic carbon-cycle observations and the need for implementing a policy-relevant carbon observing system. *Biogeosciences, 11*, 3547–3602.

CITES. (2017). Convention on International Trade in Endangered Species of Wild Fauna and Flora. Retrieved from https://cites.org/eng/disc/text.php

Clack, C. T. M., Qvist, S. A., Apt, J., Bazilian, M., Brandt, A. R., Caldeira, K., . . . Whitacre, J. F. (2017). Evaluation of a proposal for reliable low-cost grid power with 100% wind, water, and solar. *Proceedings of the National Academy of Science U.S.A., 114*, 6722–6727. doi: 10.1073/pnas.1610381114

Clapp, J., and Dauvergne, P. (2005). *Paths to a green world.* Cambridge, MA: MIT Press.

Clark, P. U., Shakun, J. D., Marcott, S. A., Mix, A. C., Eby, M., Kulp, S., . . . Plattner, G. K. (2016). Consequences of twenty-first-century policy for multi-millennial climate and sea-level change. *Nature Climate Change, 6*, 360–369.

Clark, W. C., Crutzen, P. J., and Schellnhuber, H. J. (2004). Science for global sustainability. In H. J. Schellnhuber, P. J. Crutzen, W. C. Clark, M. Claussen, and H. Held

(Eds.), *Earth system analysis for sustainability* (pp. 1–28). Cambridge, MA: MIT Press.

Clement, A. C., Burgman, R., and Norris, J. R. (2009). Observational and model evidence for positive low-level cloud feedback. *Science, 325*, 460–464.

Cohen, W. B., and Goward, S. N. (2004). Landsat's role in ecological applications of remote sensing. *BioScience, 54*, 535–545.

Cohen, W. B., Spies, T. A., and Fiorella, M. (1995). Estimating the age and structure of forests in a multi-ownership landscape of western Oregon, U.S.A. *International Journal of Remote Sensing, 16*, 721–746.

Colbourn, G., Ridgwell, A., and Lenton, T. M. (2015). The time scale of the silicate weathering negative feedback on atmospheric CO_2. *Global Biogeochemical Cycles, 29*, 583–596.

Conway Morris, S. (2015). *The runes of evolution: How the universe became self-aware.* West Conshohocken, PA: Templeton Press.

Costa, M. H., and Foley, J. A. (2000). Combined effects of deforestation and doubled CO_2 concentrations on the climate of Amazonia. *Journal of Climate, 13*, 18–34.

Cox, P. M., Betts, R. A., Jones, C. D., Spall, S. A., and Totterdell, I. J. (2000). Acceleration of global warming due to carbon cycle feedbacks in a coupled climate model. *Nature, 408*, 184–187.

Cronin, T. M. (2009). *Paleoclimates: Understanding climate change past and present.* New York, NY: Columbia University Press.

Crimmins, S. M., Dobrowski, S. Z., Greenberg, J. A., Abatzoglou, J. T., and Mynsberge, A. R. (2011). Changes in climatic water balance drive downhill shifts in plant species' optimum elevations. *Science, 331*, 324–327.

Crutzen, P. (2002). Geology of mankind. *Nature, 415*, 23.

Crutzen, P. J., and Stoermer, E. F. (2000). The Anthropocene. *Global Change Newsletter, 41*, 17–18.

Cui, Y., and Kump, L. R. (2015). Global warming and the end-Permian extinction event: Proxy and modeling perspectives. *Earth-Science Reviews, 149*, 5–22.

Cumming, G. S., Cumming, D. H. M., and Redman, C. L. (2006). Scale mismatches in social-ecological systems: Causes, consequences, and solutions. *Ecology and Society, 11* (14).

Daily, G. C. (Ed.). (1997). *Nature's services: Societal dependence on natural ecosystems.* Washington, DC: Island Press.

Daily, G. C., and Matson, P. A. (2008) Ecosystem services: From theory to implementation. *Proceedings of the National Academy of Sciences U.S.A., 105*, 9455–9456.

Dale, T., and Carter, V. G. (1955). *Topsoil and civilization.* Norman, OK: University of Oklahoma Press.

Datla, R., Weinreb, M., Rice, J., Johnson, B. C., Shirley, E., and Cao, C. Y. (2014). Optical passive sensor calibration for satellite remote sensing and the legacy of NOAA and NIST cooperation. *Journal of Research of the National Institute of Standards and Technology, 119*, 235–255.

Davies, P. C. W. (2008). Fitness and the cosmic environment. In J. D. Barrow, S. C. Morris, S. J. Freeland, and C. L. Harper, Jr. (Eds.), *Fitness of the cosmos for life: Biochemistry and fine-tuning* (pp. 97–113). Cambridge: Cambridge University Press.

Dawkins, R. (1999). *The extended phenotype*. Oxford: Oxford University Press.

DAYMET. (2017). Daymet V3: Daily surface weather and climatological summaries. Retrieved from https://daymet.ornl.gov/

DeConto, R. M., and Pollard, D. (2003). Rapid Cenozoic glaciation of Antarctica induced by declining atmospheric CO_2. *Nature, 421*, 245–249.

De Duve, C. (2002). *Life evolving: Molecules, mind, and meaning*. Oxford: Oxford University Press.

DeFries, R. S., Ellis, E. C., Chapin, F. S., Matson, P. A., Turner, B. L., Agrawal, A., . . . Syvitski, J. (2012). Planetary opportunities: A social contract for global change science to contribute to a sustainable future. *Bioscience, 62*, 603–606.

DeFries, R., and Eshleman, N. K. (2004). Land-use change and hydrologic processes: A major focus for the future. *Hydrological Processes, 18*, 2183–2186.

DeFries, R., and Nagendra, H. (2017). Ecosystem management as a wicked problem. *Science, 356*, 265–270.

Delluchi, M. A., and Jacobson, M. Z. (2011). Providing all global energy with wind, water, and solar power, Part II: Reliability, system and transmission costs, and policies. *Energy Policy, 39*, 1170–1190.

Devall, W., and Sessions, G. (1985). *Deep ecology: Living as if nature mattered*. Salt Lake City, UT: Peregrine Smith.

Diamond, J. (2005). *Collapse*. New York: Viking.

Diamond, J. (2011). *Collapse: How societies fail or succeed* (Rev. ed.). London: Penguin Group (USA).

Diaz, R. J., and Rosenberg, R. (2008) Spreading dead zones and consequences for marine ecosystems. *Science, 321*, 926–929.

Dietsch, A. M., Teel, T. L., and Manfredo, M. J. (2016). Social values and biodiversity conservation in a dynamic world. *Conservation Biology, 30*, 1212–1221.

Dietz, T., Ostrom, E., and Stern, P. C. (2003). The struggle to govern the commons. *Science, 302*, 1907–1912.

Dimitrov, R. S. (2016). The Paris agreement on climate change: Behind closed doors. *Global Environmental Politics, 16*, 1–11.

Dinar, S. (Ed.). (2011). *Beyond resource wars: Scarcity, environmental degradation, and international cooperation*. Cambridge, MA: MIT Press.

Doerr, S. H., and Santin, C. (2016). Global trends in wildfire and its impacts: Perceptions versus realities in a changing world. *Philosophical Transactions of the Royal Society B–Biological Sciences, 371*. doi: 10.1098/rstb.2015.0345

Dregne, H. E. (1991). Arid land degradation—a result of mismanagement. *Geotimes, 36*, 19–21.

Dybas, C. L. (2005). Dead zones spreading in world oceans. *Bioscience, 55*, 552–557.

Eckersley, R. (2004). *The green state: Rethinking democracy and sovereignty*. Cambridge, MA: MIT Press.

Economist. (2013). Dams in the Amazon: The rights and wrongs of Belo Monte. Retrieved from http://www.economist.com/news/americas/21577073-having-spent-heavily-make-worlds-third-biggest-hydroelectric-project-greener-brazil.

Economist. (2016a). The state of the world: Better and better. Retrieved from http://www.economist.com/news/books-and-arts/21706231-human-life-has-improved-many-ways-both-recently-according-swedish-economic.

Economist. (2016b). Dams in the Amazon: Not in my valley. Retrieved from http://www.economist.com/news/americas/21709569-hydropower-not-reliable-people-thought-new-ways-generate-electricity-are-becoming.

Edwards, M., and Richardson, A. J. (2004). Impact of climate change on marine pelagic phenology and trophic mismatch. *Nature, 430*, 881–884.

Ehrenfeld, J. G. (2010). Ecosystem consequences of biological invasions. In D. J. Futuyma, H. B. Shafer, and D. Simberloff (Eds.), *Annual review of ecology, evolution, and systematics* (pp. 59–80). Palo Alto, CA: Annual Reviews.

Ehrlich, P. R., and Ehrlich, A. H. (2009). *The dominant animal: Human evolution and the environment*. Washington, DC: Island Press.

Ehrlich, P. R., and Ehrlich, A. H. (2012). Can collapse of global civilization be avoided? *Proceedings of the Royal Society B–Biological Sciences, 280*. doi: 1098/rspb.2012.2845

Ellis, E. C. (2015). Ecology in an anthropogenic biosphere. *Ecological Monographs, 85*, 287–331

Ellis, E. C., Goldewijk, K. K., Siebert, S., Lightman, D., and Ramankutty, N. (2010). Anthropogenic transformation of the biomes, 1700 to 2000. *Global Ecology and Biogeography, 19*, 589–606.

Ellison, D., Morris, C. E., Locatelli, B., Sheil, D., Cohen, J., Murdiyarso, D., . . . Sullivan, C. A. (2017). Trees, forests and water: Cool insights for a hot world. *Global Environmental Change: Human and Policy Dimensions, 43*, 51–61.

Eltahir, E. A. B., and Bras, R. L. (1994). Precipitation recycling in the Amazon Basin. *Quarterly Journal of the Royal Meteorological Society, 120*, 861–880.

EOSDIS. (2017). EOSDIS distributed active archive centers (DAACs). Retrieved from https://earthdata.nasa.gov/about/daacs.

ESA. (2017). Copernicus: Observing the Earth. European Space Agency. Retrieved from http://www.esa.int/Our_Activities/Observing_the_Earth/Copernicus/Overview4.

Escobar, A. (2001). Culture sits in places: Reflections on globalism and subaltern strategies of localization. *Political Geography, 20*, 139–174.

Famiglietti, J. S., and Rodell, M. (2013). *Water in the balance. Science, 340*, 1300–1301.

FAO. (2014). World review of fisheries and aquaculture. Food and Agriculture Organization of the United Nations. Retrieved from http://www.fao.org/docrep/016/i2727e/i2727e01.pdf

Feely, R. A., Doney, S. C., and Cooley, S. R. (2009) Ocean acidification: Present conditions and future changes in a high-CO_2 world. *Oceanography, 22*, 36–47.

Fei, S., Desprez, J. M., Potter, K. M., Insu, J., Knott, J. A., and Sowalt, C. M. (2017). Divergence of species responses to climate change. *Science Advances, 3.* doi: 10.1126/sciadv.1603055

Field, C. B., Behrenfeld, M. J., Randerson, J. T., and Falkowski, P. (1998). Primary production of the biosphere: Integrating terrestrial and oceanic components. *Science, 281,* 237–240.

Fischer, A. G. (1984). The two Phanerozoic cycles. In A. G. Berggren and J. A. Van Couvering (Eds.), *Castastrophes and Earth history: The new uniformitarianism* (pp. 129–150). Princeton, NJ: Princeton University Press.

Fisher, R., and Freudenburg, W. R. (2001). Ecological modernization and its critics: Assessing the past and looking toward the future. *Society and Natural Resources, 14,* 701–709.

Fletcher, B. J., Brentnall, S. J., Anderson, C. W., Berner, R. A., and Beerling, D. J. (2008). Atmospheric carbon dioxide linked with Mesozoic and early Cenozoic climate change. *Nature Geoscience, 1,* 43–48.

Floudas, D., Binder, M., Riley, R., Barry, K., Blanchette, R. A., Henrissat, B., . . . Hibbett, D. S. (2012). The Paleozoic origin of enzymatic lignin decomposition reconstructed from 31 fungal genomes. *Science, 336,* 1715–1719.

FLUXNET. (2017). Retrieved from http://fluxnet.ornl.gov/

Foley, J. A., Monfreda, C., Ramankutty, N., and Zaks, D. (2007). Our share of the planetary pie. *Proceedings of the National Academy of Sciences U.S.A., 104,* 12585–12586.

Folke, C., Hahn, T., Olsson, P., and Norberg, J. (2005). Adaptive governance of social-ecological systems. *Annual Review of Environment and Resources, 30,* 441–473.

Folke, C., Jansson, A., Rockstrom, J., Olsson, P., Carpenter, S. R., Chapin, F. S., . . . Westley, F. (2011). Reconnecting to the biosphere. *Ambio, 40,* 719–738

Forkel, M., Carvalhais, N., Rodenbeck, C., Keeling, R., Heimann, M., Thonicke, K., . . . Reichstein, M. (2016). Enhanced seasonal CO_2 exchange caused by amplified plant productivity in northern ecosystems. *Science, 351,* 696–699.

Fournier, V. (2008). Escaping from the economy: The politics of degrowth. *International Journal of Sociology and Social Policy, 28,* 528 – 545.

FRA. (2010). Global forest resources assessment 2010. United Nations Food and Agriculture Organization. Retrieved from http://www.fao.org/forestry/fra/fra2010/en/

Frakes, L. A., Francis, J. E., and Syktus, J. (1992). *Climate modes of the Phanerozoic.* Cambridge: Cambridge University Press.

Franck, S., Bounama, C., and von Bloh, W. (2006). Causes and timing of future biosphere extinctions. *Biogeosciences, 3,* 85–92.

Frankenberg, C., O'Dell, C., Berry, J., Guanter, L., Joiner, J., Kohler, P., . . . Taylor, T. E. (2014). Prospects for chlorophyll fluorescence remote sensing from the Orbiting Carbon Observatory-2. *Remote Sensing of Environment, 147,* 1–12.

Frankl, G. (2003). *The social history of the unconscious.* London: Open Gate.

Franks, P. J., Royer, D. L., Beerling, D. J., Van de Water, P. K., Cantrill, D. J., Barbour, M. M., and Berry, J. A. (2014). New constraints on atmospheric CO_2 concentration for the Phanerozoic. *Geophysical Research Letters, 41,* 4685–4694.

Freeman, M. C., Wagner, G., and Zeckhouser, R. J. (2015). Climate uncertainty: When is good news bad? National Bureau of Economic Research Working Paper 20900. Retrieved from http://belfercenter.hks.harvard.edu/files/dp76_freeman-wagner-zeckhauser-2.pdf

FSC. (2017). Forest Stewardship Council. Retrieved from https://ic.fsc.org/en/certification/principles-and-criteria

Fukuyama, F. (1992). *The end of history and the last man.* New York: Free Press.

Funtowicz, S. O., and Ravetz, J. R. (1994). Uncertainty, complexity and post-normal science. *Environmental Toxicology and Chemistry, 13,* 1881–1885.

Fürst, J. J., Durand, G., Gillet-Chaulet, F., Tavard, L., Rankl, M., Braun, M., and Gagliardini, O. (2016). The safety band of Antarctic ice shelves. *Nature Climate Change, 6,* 479–482.

Galaz, V., Biermann, F., Folke, C., Nilsson, M., and Olsson, P. (2012). Global environmental governance and planetary boundaries: An introduction. *Ecological Economics, 81,* 1–3.

Galeotti, S., DeConto, R., Naish, T., Stocchi, P., Florindo, F., Pagani, M., . . . Zachos, J. C. (2016). Antarctic Ice Sheet variability across the Eocene-Oligocene boundary climate transition. *Science, 352,* 76–80.

Galli, A. M., and Fisher, D. R. (2016). Hybrid arrangements as a form of ecological modernization: The case of the US energy efficiency conservation block grants. *Sustainability, 8.* doi: 10.3390/su8010088

GCOS. (2017a). The Global Climate Observing System. Retrieved from http://www.wmo.int/pages/prog/gcos/index.php?name=about

GCOS. (2017b). Status of the Global Observing System for Climate. Retrieved from http://www.wmo.int/pages/prog/gcos/Publications/GCOS-195_en.pdf

GCP. (2017). Global Carbon Project. Retrieved from http://www.globalcarbonproject.org/carbonbudget/13/hl-compact.htm

Gell-Mann, M. (2010). Transformations of the twenty-first century: Transitions to greater sustainability. In J. Schellnhuber, M. Molina, N. Stern, V. Huber, and A. Kadner (Eds.), *Global sustainability: A Nobel cause* (pp. 1–18). Cambridge: Cambridge University Press.

GEO. (2017). Group on Earth Observations. Retrieved from http://www.earthobservations.org/wigeo.php

Giampietro, M., Aspinall, R. J., Ramos-Martin, J., and Bukkens, S. G. F. (Eds.). (2014). *Resource accounting for sustainability assessment: The nexus between energy, food, water and land use.* London: Routledge.

Gibbs, H. K., Rausch, L., Munger, J., Schelly, I., Morton, D. C., Noojipady, P., . . . Walker, N. F. (2015). Brazil's soy moratorium. *Science, 347,* 377–378.

Gibbs, H. K., and Salmon, J. M. (2015). Mapping the world's degraded lands. *Applied Geography, 57,* 12–21.

Giddens, A. (1991). *The consequences of modernity.* Cambridge: Polity Press.

Gillard, A. (1969) On terminology of biosphere and ecosphere. *Nature, 223,* 500–501.

Glaser, M., Ratter, B. M. W., Krause, G., and Welp, M. (2012). New approaches to the analysis of human–nature relations. In M. Glaser, G. Krause, B. M. W. Ratter, and M. Welp (Eds.), *Human–nature interactions in the Anthropocene: Potentials of social–ecological systems analysis* (pp. 3–12). London: Routledge.

Global Fishing Watch. (2017). Retrieved from http://globalfishingwatch.org/

Global Forest Watch. (2017). Retrieved from http://www.globalforestwatch.org/

GLOSS. (2017). The Global Sea Level Observing System. Retrieved from http://www.gloss-sealevel.org/

Glover, L. (2006). *Postmodern climate change.* London: Routledge.

GOES. (2017). Geostationary Operational Environmental Satellite. Retrieved from http://noaasis.noaa.gov/NOAASIS/ml/genlsatl.html

Goodie, A. (2006). *The human impact on the natural environment.* Hoboken, NJ: Blackwell.

GOOS. (2017). Global Ocean Observatory System. Retrieved https://www.ncdc.noaa.gov/gosic/global-ocean-observing-system-goos

Gough, D. O. (1981). Solar interior structure and luminosity variations. *Solar Physics, 74,* 21–34.

Gough, N. (2002). Thinking/acting locally/globally: Western science and environmental education in a global knowledge economy. *International Journal of Science Education, 24,* 1217–1237.

Gower, S. T., Kucharik, C. J., and Norman, J. M. (1999). Direct and indirect estimation of leaf area index, f(APAR), and net primary production of terrestrial ecosystems. *Remote Sensing of Environment, 70,* 29–51.

GPMM. (2017). Global Precipitation Measurement Mission. Retrieved from http://www.nasa.gov/sites/default/files/files/GPM_Mission_Brochure.pdf

GRACE. (2017). Gravity Recovery and Climate Experiment. Retrieved from http://www.nasa.gov/mission_pages/Grace/#.U0LkJPldXmc.

Graesser, J., Aide, T. M., Grau, H. R., and Ramankutty, N. (2015). Cropland/pastureland dynamics and the slowdown of deforestation in Latin America. *Environmental Research Letters, 10.* doi: 10.1088/1748–9326/10/3/034017

Graham, F. (2013). Earth's plant life glows green in hi-res space map. *New Scientist.* Retrieved from https://www.newscientist.com/article/dn23738-earths-plant-life-glows-green-in-hi-res-space-map/#.Uct35PnVCWY

Gregor, C. B., Garrels, R. M., Mackenzie, F. T., and Maynard, J. B. (1988). *Chemical cycles in the evolution of the Earth.* New York: Wiley.

Grimm, N. B., Faeth, S. H., Golubiewski, N. E., Redman, C. L., Wu, J. G., Bai, X. M., and Briggs, J. M. (2008). Global change and the ecology of cities. *Science, 319,* 756–760.

Grinspoon, D. (2016). *Earth in human hands.* New York: Grand Central Publishing.

Grubb, P. J. (1977). Maintenance of species-richness in plant communities—importance of regeneration niche. *Biological Reviews of the Cambridge Philosophical Society, 52,* 107–145.

GTN-R. (2017). Global Terrestrial Network for River Discharge (GTN-R). Retrieved from http://www.bafg.de/GRDC/EN/04_spcldtbss/44_GTNR/gtnr_node.html

Guillette, L. J., and Iguchi, T. (2012). Life in a contaminated world. *Science, 337,* 1614–1615.

Gunderson, L. H., and Holling, C. H. (2002). *Panarchy: Understanding transformations in natural systems.* Washington, DC: Island Press.

Haberl, H., Erb, K. H., Krausmann, F., Gaube, V., Bondeau, A., Plutzar, C., . . . Fischer-Kowalski, M. (2007). Quantifying and mapping the human appropriation of net primary production in earth's terrestrial ecosystems. *Proceedings of the National Academy of Sciences U.S.A., 104,* 12942–12945.

Hackmann, H., Moser, S. C., and St. Clair, A. L. (2014). The social heart of global environmental change. *Nature Climate Change, 4,* 653–655.

Haff, P. K. (2014). Technology as a geological phenomenon: Implications for human well-being. In C. N. Waters, J. A. Zalasiewicz, M. W. Williams, M. A. Ellis, and A. M. Snelling (Eds.), *A stratigraphical basis for the Anthropocene* (Special Publication 395:301–309). London: Geological Society.

Hall-Spencer, J. M., Rodolfo-Metalpa, R., Martin, S., Ransome, E., Fine, M., Turner, S. M., . . . Buia, M. C. (2008). Volcanic carbon dioxide vents show ecosystem effects of ocean acidification. *Nature, 454,* 96–99.

Halpern, B. S., Walbridge, S., Selkoe, K. A., Kappel, C. V., Micheli, F., D'Agrosa, C., . . . Watson, R. (2008). A global map of human impact on marine ecosystems. *Science, 319,* 948–952.

Hansen, J., Kharecha, P., Sato, M., Masson-Delmotte, V., Ackerman, F., Beerling, D. J., . . . Zachos, J. C. (2013). Assessing "dangerous climate change": Required reduction of carbon emissions to protect young people, future generations and nature. *Plos One, 8,* 26. doi: 10.1371/journal.pone.0081648

Hansen, M. C., Potapov, P. V., Moore, R., Hancher, M., Turubanova, S. A., Tyukavina, A., . . . Townshend, J. R. (2013). High-resolution global maps of 21st-century forest cover change. *Science, 342,* 850–853.

Hardin, G. (1968) The tragedy of the commons. *Science, 162,* 1243–1248.

Harvey, D. (1989). *The conditions of postmodernity.* Hoboken, NJ: Blackwell. Retrieved from https://libcom.org/files/David%20Harvey%20-%20The%20Condition%20of%20Postmodernity.pdf

Hastings, D. A., and Emery, W. J. (1992). The Advanced Very High Resolution Radiometer (AVHRR)—A brief reference guide. *Photogrammetric Engineering and Remote Sensing, 58,* 1183–1188.

Hauer, M. E., Evans, J. M., and Mishra, D. R. (2016). Millions projected to be at risk from sea-level rise in the continental United States. *Nature Climate Change, 6,* 691–695.

Hawkins, E., Smith, R. S., Allison, L. C., Gregory, J. M., Woollings, T. J., Pohlmann, H., and de Cuevas, B. (2011). Bistability of the Atlantic overturning circulation in a global climate model and links to ocean freshwater transport. *Geophysical Research Letters, 38.* doi: 10.1029/2011gl047208

Hays, G. C., Richardson, A. J., and Robinson, C. (2005) Climate change and marine plankton. *Trends in Ecology and Evolution, 20,* 337–344.

Haywood, A. M., Ridgwell, A., Lunt, D. J., Hill, D. J., Pound, M. J., Dowsett, H. J., . . . Williams, M. (2011). Are there pre-Quaternary geological analogues for a future greenhouse warming? *Philosophical Transactions of the Royal Society A–Mathematical Physical and Engineering Sciences, 369,* 933–956.

Hibbard, K. A., Crutzen, P., Lambin, E. F., Liverman, D., Mantua, N. J., McNeill, J. R., . . . Steffen, N. (2007). The Great Acceleration. In R. Costanza, L. J. Graumlich, and W. Steffen (Eds.), *Sustainability or collapse? An integrated history and future of people on Earth* (pp. 341–378). Cambridge, MA: MIT Press.

Hite, A. B., Roberts, J. T., and Chorev, N. (2015). Globalization and development. In J. T. Roberts, A. Hite, and N. Chorev (Eds.), *The globalization and development reader: Perspectives on development and global change* (pp. 1–17). Hoboken, NJ: Wiley-Blackwell.

Hoegh-Guldberg, O., and Bruno, J. F. (2010). The impact of climate change on the world's marine ecosystems. *Science, 328,* 1523–1528.

Hofgaard, A., Tommervik, H., Rees, G., and Hanssen, F. (2013). Latitudinal forest advance in northernmost Norway since the early 20th century. *Journal of Biogeography, 40,* 938–949.

Hopewell, K. (2016). *Breaking the WTO.* Stanford, CA: Stanford University Press.

Houghton, R. A., House, J. I., Pongratz, J., van der Werf, G. R., DeFries, R. S., Hansen, M. C., . . . Ramankutty, N. (2012). Carbon emissions from land use and land-cover change. *Biogeosciences, 9,* 5125–5142.

Hsu, A., and Miao, W. (2014). Soil pollution in China still a state secret [Web log post]. Retrieved from http://blogs.scientificamerican.com/guest-blog/soil-pollution-in -china-still-a-state-secret-infographic/

Huber, J. (2008). Pioneer countries and the global diffusion of environmental innovations: Theses from the viewpoint of ecological modernisation theory. *Global Environmental Change: Human and Policy Dimensions, 18,* 360–367.

Huntington, T. G. (2006). Evidence for intensification of the global water cycle: Review and synthesis. *Journal of Hydrology, 319,* 83–95.

Hurtt, G. C., Chini, L. P., Frolking, S., Betts, R. A., Feddema, J., Fischer, G., . . . Wang, Y. P. (2011). Harmonization of land-use scenarios for the period 1500–2100: 600 years of global gridded annual land-use transitions, wood harvest, and resulting secondary lands. *Climatic Change, 109,* 117–161.

Hutchinson, G. E. (1970). The biosphere. *Scientific American, 223,* 45–53.

Huxley, J. (1958). Introduction to *The Phenomenon of Man.* In P. R. Sampson and D. Pitt (Eds.), *The biosphere and noösphere reader* (pp. 80–85). London: Routledge.

IMAGE. (2017). Integrated Model to Assess the Global Environment. Retrieved from http://www.pbl.nl/en/publications/integrated-assessment-of-global-environmen tal-change-with-IMAGE-3.0

Imhoff, M. L., Bounoua, L., Ricketts, T., Loucks, C., Harriss, R., and Lawrence, W. T. (2004). Global patterns in human consumption of net primary production. *Nature, 429,* 870–873.

INFEWS. (2017). Innovations at the Nexus of Food, Energy and Water Systems. https:// www.nsf.gov/pubs/2016/nsf16524/nsf16524.htm

Inman-Narahari, F., Ostertag, R., Asner, G. P., Cordell, S., Hubbell, S. P., and Sack, L. (2014). Trade-offs in seedling growth and survival within and across tropical forest microhabitats. *Ecology and Evolution*, 4, 3755–3767.

IPBES. (2017). Intergovernmental Science-Policy Platform on Biodiversity and Ecosystem Services. Retrieved from http://www.ipbes.net/about-ipbes.html

IPCC. (2005). *IPCC special report on carbon dioxide capture and storage. Prepared by Working Group III of the Intergovernmental Panel on Climate Change* [B. Metz, O. Davidson, de H. C. Coninck, M. Loos, and L. A. Meyer (Eds.)]. Cambridge: Cambridge University Press.

IPCC. (2007). Trends in the hydroxyl free radical. IPCC AR4 WG1. Retrieved from http://www.ipcc.ch/pdf/assessment-report/ar4/wg1/ar4-wg1-chapter2.pdf.

IPCC; Stocker, T. F., Qin, D., Plattner, G. K., Alexander, L. V., Allen, S. K., Bindoff, N. L., . . . Xie, S. P. (2013). Technical summary. In Stocker, T. F., Qin, D. Plattner, G. K., Tignor, M., Allen, S. K., Boschung, J., . . . Midgley (Eds.), *Climate change 2013: The physical science basis. Contribution of Working Group I to the Fifth Assessment Report of the Intergovernmental Panel on Climate Change* (pp. 33–115). Cambridge: Cambridge University Press. doi:10.1017/ CBO9781107415324.005.

IPCC. (2014a) IPCC fifth assessment report. Retrieved from https://www.ipcc.ch/ report/ar5/

IPCC. (2014b). *Climate change 2013: The physical science basis.* [T. F. Stocker, D. Qin, G. K. Plattner, M. M. B. Tignor, S. K. Allen, S. K., J. Boschung, . . . P. M. Midgeley, P. M. (Eds.)]. Retrieved from https://www.ipcc.ch/report/ar5/wg1

IPCC. (2014c). *Climate change 2013: Impacts, adaptation, and vulnerability.* [T. F. Stocker, D. Qin, G. K. Plattner, M. M. B. Tignor, S. K. Allen, S. K., J. Boschung, . . . P. M. Midgeley, P. M. (Eds.)]. Retrieved from https://www.ipcc.ch/report/ar5/wg2

IPCC. (2014d). *Climate change 2013: Mitigation of climate change.* [T. F. Stocker, D. Qin, G. K. Plattner, M. M. B. Tignor, S. K. Allen, S. K., J. Boschung, . . . P. M. Midgeley, P. M. (Eds.)]. Retrieved from https://www.ipcc.ch/report/ar5/wg3

IPCC. (2014e). Sea level change. [T. F. Stocker, D. Qin, G. K. Plattner, M. M. B. Tignor, S. K. Allen, S. K., J. Boschung, . . . P. M. Midgeley, P. M. (Eds.)]. Retrieved from https:// www.ipcc.ch/pdf/assessment-report/ar5/wg1/WG1AR5_Chapter13_FINAL.pdf

IPCC. (2015). Technical summary. Box TS.1. Retrieved from https://www.ipcc.ch/pdf/ assessment-report/ar5/wg2/WGIIAR5-TS_FINAL.pdf

IUCN. (2017). International Union for the Conservation of Nature. Retrieved from http://www.iucnredlist.org/about/summary-statistics

Ivanova, M. (2005). Assessing UNEP as anchor institution for the global environment: Lessons for the UNEO debate. Retrieved from http://www.unep.org/environmen talgovernance/Portals/8/AnchorInstitutionGlobalEnvironment.pdf

Jackson, R. B., Canadell, J. G., Le Quere, C., Andrew, R. M., Korsbakken, J. I., Peters, G. P., and Nakicenovic, N. (2016). Reaching peak emissions. *Nature Climate Change*, 6, 7–10.

Jacobson, M. Z., and Delucchi, M. A. (2009). A path to sustainable energy by 2030. *Scientific American, 301*, 58–65.

Jacobson, M. Z., and Delucchi, M. A. (2011). Providing all global energy with wind, water, and solar power, Part I: Technologies, energy resources, quantities and areas of infrastructure, and materials. *Energy Policy, 39*, 1154–1169.

Jacobson, M. Z., Delucchi, M. A., Cameron, M. A., and Frew, B. A. (2015). Low-cost solution to the grid reliability problem with 100% penetration of intermittent wind, water, and solar for all purposes. *Proceedings of the National Academy of Sciences U.S.A., 112*, 15060–15065.

Jamison, A., and Baark, E. (1999). National shades of green: Comparing the Swedish and Danish styles in ecological modernisation. *Environmental Values, 8*, 199–218.

Jänicke, M., and Jörgens, H. (2009). New approaches to environmental governance. In A. P. J. Mol, D. A. Sonnenfeld, and G. Spaargaren (Eds.), *The ecological modernization reader* (pp. 157–187). London: Routledge.

Jantz, S. M., Barker, B., Brooks, T. M., Chini, L. P., Huang, Q. Y., Moore, R. M., . . . Hurtt, G. C. (2015). Future habitat loss and extinctions driven by land-use change in biodiversity hotspots under four scenarios of climate-change mitigation. *Conservation Biology, 29*, 1122–1131.

Jenkins, C. N., and Joppa, L. (2009). Expansion of the global terrestrial protected area system. *Biological Conservation, 142*, 2166–2174.

Jetz, W., Cavender-Bares, J., Pavlick, R., Schimel, D., Davis, F. W., Asner, G. P., . . . Ustin, S. L. (2016). Monitoring plant functional diversity from space. *Nature Plants, 2*. doi: 10.1038/nplants.2016.24

Ju, J. C., and Masek, J. G. (2016). The vegetation greenness trend in Canada and US Alaska from 1984–2012 Landsat data. *Remote Sensing of Environment, 176*, 1–16.

Justice, C., Belward, A., Morisette, J., Lewis, P., Privette, J., and Baret, F. (2000). Developments in the 'validation' of satellite sensor products for the study of the land surface. *International Journal of Remote Sensing, 21*, 3383–3390.

Kates, R. W. (2010). Readings in sustainability science and technology. (CID Working Paper 213). Retrieved from https://www.hks.harvard.edu/centers/cid/publications/faculty-working-papers/cid-working-paper-no.-213

Kates, R. W., Clark, W. C., Corell, R., Hall, J. M., Jaeger, C. C., Lowe, I., . . . Svedlin, U. (2001). Environment and development—Sustainability science. *Science, 292*, 641–642.

Kauffman, S. (1993). *The origins of order.* Oxford: Oxford University Press.

Kauffman, S. (1995). *At home in the universe.* Oxford: Oxford University Press.

Kauffman, S. (2008). *Reinventing the sacred.* New York: Basic Books.

Keeling, C. D. (1998). Rewards and penalties of monitoring the earth. *Annual Review of Energy and the Environment, 23*, 25–82.

Keeling, R. F., Kortzinger, A., and Gruber, N. (2010). Ocean deoxygenation in a warming world. *Annual Review of Marine Science, 2*, 199–229.

Keenan, R. J., Reams, G. A., Achard, F., de Freitas, J. V., Grainger, A., and Lindquist, E. (2015). Dynamics of global forest area: Results from the FAO Global Forest Resources Assessment 2015. *Forest Ecology and Management, 352*, 9–20.

Keenan, T. F., Hollinger, D. Y., Bohrer, G., Dragoni, D., Munger, J. W., Schmid, H. P., and Richardson, A. D. (2013). Increase in forest water-use efficiency as atmospheric carbon dioxide concentrations rise. *Nature, 499,* 324–329.

Kelly, R., Chipman, M. L., Higuera, P. E., Stefanova, I., Brubaker, L. B., and Hu, F. S. (2013). Recent burning of boreal forests exceeds fire regime limits of the past 10,000 years. *Proceedings of the National Academy of Sciences U.S.A., 110,* 13055–13060.

Kemp, L. (2017). Better out than in. *Nature Climate Change.* doi: 10.1038/nclimate3309

Kennedy, R. E., Yang, Z. G., and Cohen, W. B. (2010). Detecting trends in forest disturbance and recovery using yearly Landsat time series: 1. LandTrendr—Temporal segmentation algorithms. *Remote Sensing of Environment, 114,* 2897–2910.

Kenward, R. E., Whittingham, M. J., Arampatzis, S., Manos, B. D., Hahn, T., Terry, A., . . . Rutz, C. (2011). Identifying governance strategies that effectively support ecosystem services, resource sustainability, and biodiversity. *Proceedings of the National Academy of Sciences U.S.A., 108,* 5308–5312.

Khoury, C. K., Bjorkman, A. D., Dempewolf, H., Ramirez-Villegas, J., Guarino, L., Jarvis, A., . . . Struik, P. C. (2014). Increasing homogeneity in global food supplies and the implications for food security. *Proceedings of the National Academy of Sciences U.S.A., 111,* 4001–4006.

Kiehl, J. T., and Trenberth, K. E. (1997). Earth's annual global mean energy budget. *Bulletin of the American Meteorological Society, 78,* 197–208.

Kirchner, J. W. (2003). The Gaia hypothesis: Conjectures and refutations. *Climatic Change, 58,* 21–45.

Klein, N. (2014). *This changes everything: Capitalism vs. the climate.* New York: Simon and Schuster.

Klein Goldewijk, K. (2004). Footprints from the past: Blueprint for the future. In R. DeFries, G. P. Asner, and R. Houghton (Eds.), *Ecosystems and land use change* (pp. 203–215). Washington, DC: American Geophysical Union.

Knoll, A. H., Bambach, R. K., Payne, J. L., Pruss, S., and Fischer, W. W. (2007). Paleophysiology and end-Permian mass extinction. *Earth and Planetary Science Letters, 256,* 295–313.

Kort, E. A., Frankenberg, C., Costigan, K. R., Lindenmaier, R., Dubey, M. K., and Wunch, D. (2014). Four corners: The largest US methane anomaly viewed from space. *Geophysical Research Letters, 41,* 6898–6903.

Krausmann, F., Erb, K. H., Gingrich, S., Haberl, H., Bondeau, A., Gaube, V., . . . Searchinger, T. D. (2013). Global human appropriation of net primary production doubled in the 20th century. *Proceedings of the National Academy of Sciences U.S.A., 110,* 10324–10329.

Kreisberg, J. C. (1995), A globe, clothing itself with a brain. *Wired, 3.06.* Retrieved from https://www.wired.com/1995/06/teilhard/.

Kriegler, E., Hall, J. W., Held, H., Dawson, R., and Schellnhuber, H. J. (2009). Imprecise probability assessment of tipping points in the climate system. *Proceedings of the National Academy of Sciences U.S.A., 106,* 5041–5046.

Kulcsar, L. J. (2016). Population and development. In G. Hooks (Ed.), *The sociology of development handbook* (pp. 48–68). Oakland: University of California Press.

Kump, L. R. (2011). The last great global warming. *Scientific American, 305,* 56–61.

Kump, L. R., and Barley, M. E. (2007). Increased subaerial volcanism and the rise of atmospheric oxygen 2.5 billion years ago. *Nature, 448,* 1033–1036.

Kump, L. R., Kasting, J. F., and Crane, R. G. (2010). *The Earth system.* Upper Saddle River, NJ: Prentice Hall.

Kump, L. R., Pavlov, A., and Arthur, M. A. (2005). Massive release of hydrogen sulfide to the surface ocean and atmosphere during intervals of oceanic anoxia. *Geology, 33,* 397–400.

Laland, K. N., Uller, T., Feldman, M. W., Sterelny, K., Muller, G. B., Moczek, A., . . . Odling-Smee, J. (2015). The extended evolutionary synthesis: Its structure, assumptions and predictions. *Proceedings of the Royal Society B–Biological Sciences, 282,* 20151019. doi: 10.1098/rspb.2015.1019.

Lambin, E. F., Geist, H., and Rindfuss, R. R. (2006). Introduction: Local processes with global impacts. In E. F. Lambin and H. Geist (Eds.), *Land-use and land-cover change* (pp. 1–8). Berlin: Springer-Verlag.

Lamphere, J. A., and Shefner, J. (2015). Stimulating the green economy: Discourses, cooption, and state change. In J. Shefner (Ed.), *Current perspectives in social theory* (pp. 101–124). Bingley, UK: Emerald Group.

Lant, C. L., Ruhl, J. B., and Kraft, S. E. (2008). The tragedy of ecosystem services. *Bioscience, 58,* 969–974.

Larsen, J. (2004). The sixth great extinction: A status report. Retrieved from http://www.earth-policy.org/Updates/Update35.htm

Lashof, D. A., and Ahuja, D. R. (1990). Relative contributions of greenhouse gas emissions to global warming. *Nature, 344,* 529–531.

Lashof, D. A., DeAngelo, B. J., Saleska, S. R., and Harte, J. (1997). Terrestrial ecosystem feedbacks to global climate change. *Annual Review of Energy and the Environment, 22,* 75–118.

Laskar, J., Joutel, F., and Robutel, P. (1993). Stabilization of Earth's obliquity by the moon. *Nature, 361,* 615–617.

Lauer, D. T., Morain, S. A., and Salomonson, V. V. (1997). The Landsat program: Its origins, evolution, and impacts. *Photogrammetric Engineering and Remote Sensing, 63,* 831–838.

Lautensach, A. (2008). Environmental ethics and the Gaia idea. In, *The Encyclopedia of the Earth.* Retrieved from http://editors.eol.org/eoearth/wiki/Environmental_ethics_and_the_Gaia_theory

Lawrence, D. M., and Fisher, R. (2013). The community land model philosophy: Model development and science applications. *iLEAPS Newsletter, 13,* 16–19.

LCC. (2016). Landscape Conservation Cooperative. Retrieved from https://lccnetwork.org/

Le Quere, C., Andrew, R. M., Canadell, J. G., Sitch, S., Korsbakken, J. I., Peters, G. P., . . . Zaehle, S. (2016). Global carbon budget 2016. *Earth System Science Data*, *8*, 605–649.

Le Treut, H., and Somerville, R. (Coords.). (2007). *Historical overview of climate change science*. Cambridge: Cambridge University Press.

Leakey, R. E. (author), and Lewin, R. (contrib.). (1995). *The sixth extinction: Patterns of life and the future of humankind*. New York: Doubleday.

Leemans, R. (2016). The lessons learned from shifting from global-change research programmes to transdisciplinary sustainability science. *Current Opinion in Environmental Sustainability*, *19*, 103–110.

Lefsky, M. (2010). A global forest canopy height map from the Moderate Resolution Imaging Spectroradiometer and the Geoscience Laser Altimeter System. *Geophysical Research Letters*, *37*, L15401.

Lelieveld, J., Dentener, F. J., Peters, W., and Krol, M. C. (2004). On the role of hydroxyl radicals in the self-cleansing capacity of the troposphere. *Atmospheric Chemistry and Physics*, *4*, 2337–2344.

Lenton, T. M. (1998). Gaia and natural selection. *Nature*, *394*, 439–447.

Lenton, T. M. (2002). Testing Gaia: The effect of life on Earth's habitability and regulation. *Climatic Change*, *52*, 409–422.

Lenton, T. M. (2016). *Earth system science: A very short introduction*. Oxford: Oxford University Press.

Lenton, T. M., Held, H., Kriegler, E., Hall, J. W., Lucht, W., Rahmstorf, S., and Schnellhuber, H. J. (2008). Tipping elements in the Earth's climate system. *Proceedings of the National Academy of Sciences U.S.A.*, *105*, 1786–1793.

Lenton, T., and Watson, A. (2011). *Revolutions that made Earth*. Oxford: Oxford University Press.

Leroy, P., and van Tatenhove, J. (2000). Political modernization theory and environmental politics. In G. Spaargaen, A. P. J. Mol, and F. H. Buttel (Eds.), *Environment and global modernity* (pp. 187–208). Thousand Oaks, CA: Sage.

Levasseur, M. (2011). Ocean science: If Gaia could talk. *Nature Geoscience*, *4*, 351–352.

Levi, M. (2009). Copenhagen's inconvenient truth: How to salvage the climate conference. *Foreign Affairs*, *88*, 92–103.

Levin, S. A. (1998). Ecosystems and the biosphere as complex adaptive systems. *Ecosystems*, *1*, 431–436.

Lewis-Williams, D. (2002). *The mind in the cave: Consciousness and the origins of art*. London: Thames and Hudson.

Lightfoot, K. G., and Cuthrell, R. Q. (2015). Anthropogenic burning and the Anthropocene in late-Holocene California. *Holocene*, *25*, 1581–1587.

Likens, G. E., Bormann, F. H., Pierce, R. S., and Reiners, W. A. (1978). Recovery of a deforested ecosystem. *Science*, *199*, 492–496.

Lindenmayer, D., Hobbs, R. J., Montague-Drake, R., Alexandra, J., Bennett, A., Burgman, M., . . . Zavaleta, E. (2008). A checklist for ecological management of landscapes for conservation. *Ecology Letters*, *11*, 78–91.

Lipschutz, R. D. (1996). *Global civil society and global environmental governance*. Albany: State University of New York Press.

Littell, J. S., McKenzie, D., Peterson, D. L., and Westerling, A. L. (2009). Climate and wildfire area burned in western U.S. ecoprovinces, 1916–2003. *Ecological Applications, 19*, 1003–1021.

Liu, J. G., Dietz, T., Carpenter, S. R., Alberti, M., Folke, C., Moran, E., . . . Taylor, W. W. (2007a). Complexity of coupled human and natural systems. *Science, 317*, 1513–1516.

Liu, J. G., Dietz, T., Carpenter, S. R., Folke, C., Alberti, M., Redman, C. L., . . . Provencher, W. (2007b). Coupled human and natural systems. *Ambio, 36*, 639–649.

Lomolino, M. V., Channell, R., Perault, D. R., and Smith, G. A. (2001). Downsizing nature: Anthropogenic dwarfing of species and ecosystems. In J. L. Lockwood and M. L. McKinney (Eds.), *Biotic homogenization* (pp. 223–244). Dordrecht, Netherlands: Kluwer.

Loreau, M., Naeem, S., Inchausti, P., Bengtsson, J., Grime, J. P., Hector, A., . . . Wardle, D. A. (2001). Biodiversity and ecosystem functioning: Current knowledge and future challenges. *Science, 294*, 804–808.

Lotman, Y. M. (1990). *Universe of the mind: A semiotic theory of culture* (Trans. A. Shukman). London Tauris.

Lourantou, A., Lavric, J. V., Kohler, P., Barnola, J. M., Paillard, D., Michel, E., . . . Chappellaz, J. (2010). Constraint of the CO_2 rise by new atmospheric carbon isotopic measurements during the last deglaciation. *Global Biogeochemical Cycles, 24*, GB15.

Lovbrand, E., Stripple, J., and Wiman, B. (2009) Earth System governmentality: Reflections on science in the Anthropocene. *Global Environmental Change: Human and Policy Dimensions, 19*, 7–13.

Lovelock, J. (1979). *Gaia, a new look at life on Earth*. Oxford: Oxford University Press.

Lovelock, J. E., and Margulis, L. (1974). Atmospheric homeostasis by and for biosphere—the Gaia hypothesis. *Tellus, 26*, 2–10.

Lovelock, J. E., and Watson, A. J. (1981). The regulation of carbon dioxide and climate: Gaia or geochemistry. *Planetary and Space Science, 30*, 795–802.

LP DAAC. (2017). Land Processes Distributed Active Archive Center. Retrieved from https://lpdaac.usgs.gov/

Lubchenco, J. (1998). Entering the century of the environment: A new social contract for science. *Science, 279*, 491–497.

Lubchenco, J., Barner, A. K., Cerny-Chipman, E. B., and Reimer, J. N. (2015). Sustainability rooted in science. *Nature Geoscience, 8*, 741–745.

Lubchenco, J., Cerny-Chipman, E. B., Reimer, J. N., and Levin, S. A. (2016). The right incentives enable ocean sustainability successes and provide hope for the future. *Proceedings of the National Academy of Sciences U.S.A., 113*, 14507–14514.

Lubchenco, J., and Grorud-Colvert, K. (2015). Making waves: The science and politics of ocean protection. *Science, 350*, 382–383.

Lubchenco, J., and Petes, L. E. (2010). The interconnected biosphere: Science at the ocean's tipping points. *Oceanography, 23*, 115–129.

Lubin, D. A., and Esty, D. C. (2010). The sustainability imperative. *Harvard Business Review, 88*, 42–50.

Lueck, D. (1995). Property-rights and economics of wildlife institutions. *Natural Resources Journal, 35*, 625–670.

Luo, Y. Q., Randerson, J. T., Abramowitz, G., Bacour, C., Blyth, E., Carvalhais, N., . . . Zhou, X. H. (2012). A framework for benchmarking land models. *Biogeosciences, 9*, 3857–3874.

Lutz, W., and Samir, K. C. (2010). Dimensions of global population projections: What do we know about future population trends and structures? *Philosophical Transactions of the Royal Society B–Biological Sciences, 365*, 2779–2791.

Malhi, Y. (2014). The metabolism of a human-dominated planet. In I. Goldin (Ed.), *Is the planet full?* (pp. 142–163). Oxford: Oxford University Press.

Malhi, Y., Aragao, L., Galbraith, D., Huntingford, C., Fisher, R., Zelazowski, P., . . . Meir, P. (2009). Exploring the likelihood and mechanism of a climate-change-induced dieback of the Amazon rainforest. *Proceedings of the National Academy of Sciences U.S.A., 106*, 20610–20615.

Malm, A., and Hornborg, A. (2014). The geology of mankind? A critique of the Anthropocene narrative. *Anthropocene Review, 1*, 62–69.

Margulis, L. (1970). *The origin of eukaryotic cells: Evidence and research implications for a theory of the origin and evolution of microbial, plant and animal cells on the Precambrian Earth*. New Haven, CT: Yale University Press.

Margulis, L. (2006). Biosphere technologies and the myth of individuality. Retrieved from http://www.eoearth.org/article/The_Future_of_Human_Nature%3A_A_Sy mposium_on_the_Promises_and_Challenges_of_the_Revolutions_in_Genomics _and_Computer_Science_%28Conference%29%3A_Session_Five

Marinov, I., Follows, M., Gnanadesikan, A., Sarmiento, J. L., and Slater, R. D. (2008). How does ocean biology affect atmospheric pCO_2? Theory and models. *Journal of Geophysical Research-Oceans, 113* (C7), C07032.

Martinez, C. L. F. (2014). SETI in the light of cosmic convergent evolution. *Acta Astronautica, 104*, 341–349.

Martinez-Boti, M. A., Marino, G., Foster, G. L., Ziveri, P., Henehan, M. J., Rae, J. W. B., . . . Vance, D. (2015). Boron isotope evidence for oceanic carbon dioxide leakage during the last deglaciation. *Nature, 518*, 219–222.

Matson, P. (2009). The sustainability transition. *Issues in Science and Technology, 25*, 39–42.

Matson, P., Clark, W. C., and Anderson, K. (2016) *Pursuing sustainability*. Princeton, NJ: Princeton University Press.

Matson, P. A., and Vitousek, P. M. (2006). Agricultural intensification: Will land spared from farming be land spared for nature? *Conservation Biology, 20*, 709–710.

Matthews, B., Narwani, A., Hausch, S., Nonaka, E., Peter, H., Yamamichi, M., . . . Turner, C. B. (2011). Toward an integration of evolutionary biology and ecosystem science. *Ecology Letters, 14*, 690–701.

Maxbauer, D. P., Royer, D. L., and LePage, B. A. (2014). High Arctic forests during the middle Eocene supported by moderate levels of atmospheric CO2. *Geology, 42,* 1027–1030.

McCarthy, J. (2015). A socioecological fix to capitalist crisis and climate change? The possibilities and limits of renewable energy. *Environment and Planning A, 47,* 2485–2502.

McCauley, D. J., Woods, P., Sullivan, B., Bergman, B., Jabonicky, C., Roan, A., . . . Worm, B. (2016). Ending hide and seek at sea. *Science, 351,* 1148–1150.

McClanahan, T. R., and Shafir, S. H. (1990). Causes and consequences of sea-urchin abundance and diversity in Kenyan coral-reef lagoons. *Oecologia, 83,* 362–370.

McGinnis, M. (1998). *Bioregionalism.* Abingdon-on-Thames, UK: Routledge.

McGuffie, K., and Henderson-Sellers, A. (2005). *A climate modeling primer* (3rd ed.). Hoboken, NJ: Wiley.

McKibben, B. (2016). A world at war. Retrieved from https://newrepublic.com/article /135684/declare-war-climate-change-mobilize-wwii

McNeill, J. R. (2000). *Something new under the sun.* New York: Norton.

MEA. (2003). Millennium Ecosystem Assessment. Retrieved from http://www .millenniumassessment.org/en/index.html

MEA. (2005). *Ecosystems and human well-being: Synthesis report.* Washington, DC: Island Press.

Meadows, D. H., Meadows, D. L., Randers, J., Behren, W. W. (1972). *Limits to growth.* New York: Universe Books.

Meinshausen, M., Smith, S. J., Calvin, K., Daniel, J. S., Kainuma, M. L. T., Lamarque, J-F., . . . van Vuuren, D. P. P. (2011). The RCP greenhouse gas concentrations and their extensions from 1765 to 2300. *Climatic Change, 109,* 213–241.

Melezhik, V. A. (2006). Multiple causes of Earth's earliest global glaciation. *Terra Nova, 18,* 130–137.

MFC. (2016). Marine Fisheries Council. Retrieved from http://www.nmfs.noaa.gov/ocs /mafac/meetings/2007_12/docs/msc_principles_and_criteria.pdf

Miles, E. L. (2009). On the increasing vulnerability of the world ocean to multiple stresses. *Annual Review of Environmental Resources, 34,* 17–41.

Miles, E. L., Elsner, M. M., Littell, J. S., Binder, L. W., and Lettenmaier, D. P. (2010). Assessing regional impacts and adaptation strategies for climate change: The Washington Climate Change Impacts Assessment. *Climatic Change, 102,* 9–27.

MODIS. (2017). Moderate Resolution Imaging Spectroradiometer. Retrieved from http://modis.gsfc.nasa.gov/

Mol, A. P. J. (2000). Globalization and environment: Between apocalypse-blindness and ecological modernization. In G. Spaargaren, A. P. J. Mol, and F. H. Buttel (Eds.), *Environment and global modernity* (pp. 121–149). Thousand Oaks, CA: Sage.

Mol, A. P. J. (2006). Environment and modernity in transitional China: Frontiers of ecological modernization. *Development and Change, 37,* 29–56.

Mol, A. P. J., and Spaargaren, G. (2004). Ecological modernization and consumption: A reply. *Society and Natural Resources, 17,* 261–265.

Mol, A. P. J., Sonnenfeld, D. A., and Spaargaren, G. (2009). *The ecological moderniza-tion reader: Environmental reform in theory and practice.* London: Routledge.

Montgomery, D. R. (2008). *Dirt: The erosion of civilizations.* Oakland: University of California Press.

Moore, K. D., and Nelson, M. P. (2011). *Moral ground: Ethical action for a planet in peril.* San Antonio, TX: Trinity University Press.

Morales-Hidalgo, D., Oswalt, S. N., and Somanathan, E. (2015). Status and trends in global primary forest, protected areas, and areas designated for conservation of bio-diversity from the Global Forest Resources Assessment 2015. *Forest Ecology and Management, 352,* 68–77.

Moriarty, P., and Honnery, D. (2012). What is the global potential for renewable energy? *Renewable and Sustainable Energy Reviews, 16,* 244–252.

MRWC. (2017). Mary's River Watershed Council. Retrieved from http://www.mrwc .org/.

Muller, C., Stehfest, E., van Minnen, J. G., Strengers, B., von Bloh, W., Beusen, A. H. W., . . . Lucht, W. (2016). Drivers and patterns of land biosphere carbon balance reversal. *Environmental Research Letters, 11,* 11–22.

Murtugudde, R. (2010). Earth system science and the second Copernican revolution. *Current Science, 98,* 1579–1583.

MWCC. (2017). Montana Watershed Coordination Council. Retrieved from http:// mtwatersheds.org/app/

Nagelkerken, I., and Connell, S. D. (2015). Global alteration of ocean ecosystem func-tioning due to increasing human CO2 emissions. *Proceedings of the National Acad-emy of Sciences U.S.A., 112,* 13272–13277.

Naidoo, R., Balmford, A., Costanza, R., Fisher, B., Green, R. E., Lehner, B., . . . Ricketts, T. H. (2008). Global mapping of ecosystem services and conservation priorities. *Proceedings of the National Academy of Sciences U.S.A., 105,* 9495–9500.

Nair, P. J., Froidevaux, L., Kuttippurath, J., Zawodny, J. M., Russell, J. M., Steinbrecht, W., . . . Anderson, J. (2015). Subtropical and midlatitude ozone trends in the strato-sphere: Implications for recovery. *Journal of Geophysical Research–Atmospheres, 120,* 7247–7257.

Nair, U. S., Lawton, R. O., Welch, R. M., and Pielke, R. A. Sr. (2003). Impact of land use on Costa Rican tropical montane cloud forests: Sensitivity of cumulus cloud field characteristics to lowland deforestation. *Journal of Geophysical Research, 108,* D7.

Natureserve. (2003). America's least wanted. *Nature Conservancy.* Retrieved from http: //www.natureserve.org/library/americasleastwanted2003.pdf

Naveh, Z. (2007). Landscape ecology and sustainability. *Landscape Ecology, 22,* 1437–1440.

NCA. (2014a). National Climate Assessment: Northwest region. Retrieved from http:// nca2014.globalchange.gov/report/regions/northwest

NCA. (2014b). National Climate Assessment: Climate change impacts in the United States. Retrieved from http://nca2014.globalchange.gov/

NCEI. (2016). National Center for Environmental Information: GCOS essential climate variable (ECV) data access matrix. Retrieved from https://www.ncdc.noaa.gov/gosic/gcos-essential-climate-variable-ecv-data-access-matrix

Neelin, J. D., Munnich, M., Su, H., Meyerson, J. E., and Holloway, C. E. (2006). Tropical drying trends in global warming models and observations. *Proceedings of the National Academy of Sciences U.S.A.*, *103*, 6110–6115.

Nelson, E., Mendoza, G., Regetz, J., Polasky, S., Tallis, H., Cameron, D. R., . . . Shaw, M. R. (2009). Modeling multiple ecosystem services, biodiversity conservation, commodity production, and tradeoffs at landscape scales. *Frontiers in Ecology and the Environment*, *7*, 4–11.

Nemani, R. R., Keeling, C. D., Hashimoto, H., Jolly, W. M., Piper, S. C., Tucker, C. J., . . . Running, S. W. (2003). Climate-driven increases in global terrestrial net primary production from 1982 to 1999. *Science*, *300*, 1560–1563.

Neumann, B., Vafeidis, A. T., Zimmermann, J., and Nicholls, R. J. (2015). Future coastal population growth and exposure to sea-level rise and coastal flooding—a global assessment. *Plos One*, *10*(6), e0131375.

Ney, S., and Verweij, M. (2015). Messy institutions for wicked problems: How to generate clumsy solutions? *Environment and Planning C–Government and Policy*, *33*, 1679–1696.

Nilsson, M., and Persson, A. (2012). Can Earth system interactions be governed? Governance functions for linking climate change mitigation with land use, freshwater and biodiversity protection. *Ecological Economics*, *75*, 61–71.

NOAA. (2017a). Earth System Research Laboratory, Global Monitoring Division. Retrieved from http://www.esrl.noaa.gov/gmd/

NOAA. (2017b). El Niño theme page. Retrieved from http://www.pmel.noaa.gov/elnino/

Noble, D. (2008). Genes and causation. *Philosophical Transactions of the Royal Society A–Mathematical Physical and Engineering Sciences*, *366*, 3001–3015.

Nordhaus, W. (2015). Climate clubs to overcome free-riding. *Issues in Science and Technology*, *31*, 27–34.

Norris, R. D., Turner, S. K., Hull, P. M., and Ridgwell, A. (2013). Marine ecosystem responses to Cenozoic global change. *Science*, *341*, 492–498.

NOWC. (2017). Network of Oregon Watersheds. Retrieved from http://www.oregonwatersheds.org/

NRC. (1999). Our common journey: A transition toward sustainability. U.S. National Research Council. Retrieved from https://www.nap.edu/catalog/9690/our-common-journey-a-transition-toward-sustainability

Nriagu, J. O. (1989). A global assessment of natural sources of atmospheric trace-metals. *Nature*, *338*, 47–49.

NSIDC. (2013). Global glacier retreat. National Snow and Ice Data Center. Retrieved from http://nsidc.org/glims/glaciermelt/index.html

NSDIC. (2017). National Snow and Ice Data Center. Retrieved from http://nsidc.org/

Obermeister, N. (2017). From dichotomy to duality: Addressing interdisciplinary epistemological barriers to inclusive knowledge governance in global environmental assessments. *Environmental Science and Policy, 67*, 80–86.

OCO. (2017). Orbital Carbon Observatory. Retrieved from http://oco.jpl.nasa.gov/

Odum, E. P. (1959). *Fundamentals of ecology*. Philadelphia: Saunders.

Oliver, T. H., Heard, M. S., Isaac, N. J. B., Roy, D. B., Procter, D., Eigenbrod, F., . . . Bullock, J. M. (2015). Biodiversity and resilience of ecosystem functions. *Trends in Ecology and Evolution, 30*, 673–684.

O'Neil, K. (2009). *The environment and international relations*. Cambridge: Cambridge University Press.

O'Neill, B. C., Jiang, L. W., and Gerland, P. (2015) Plausible reductions in future population growth and implications for the environment. *Proceedings of the National Academy of Sciences U.S.A., 112*, 506–506.

Ostrom, E. (1990). *Governing the commons: The evolution of institutions for collective action*. Cambridge: Cambridge University Press.

Ostrom, E. (2007). A diagnostic approach for going beyond panaceas. *Proceedings of the National Academy of Sciences U.S.A., 104*, 15181–15187.

Ostrom, E. (2008). The challenge of common-pool resources. *Environment, 50*, 9–20.

Ostrom, E. (2009). A general framework for analyzing sustainability of socio-ecological systems. *Science, 325*, 419–422.

Ostrom, E. (2010). Polycentric systems for coping with collective action and global environmental change. *Global Environmental Change: Human and Policy Dimensions, 20*, 550–557.

Ostrom, E., Burger, J., Field, C. B., Norgaard, R. B., and Policansky, D. (1999). Sustainability—Revisiting the commons: Local lessons, global challenges. *Science, 284*, 278–282.

Packard, R. M. (2016). *A history of global health*. Baltimore: John Hopkins University Press.

Pagani, M., Caldeira, K., Berner, R., and Beerling, D. J. (2009). The role of terrestrial plants in limiting atmospheric CO_2 decline over the past 24 million years. *Nature, 460*, 85–89.

Pagani, M., Liu, Z. H., LaRiviere, J., and Ravelo, A. C. (2010). High Earth-system climate sensitivity determined from Pliocene carbon dioxide concentrations. *Nature Geoscience, 3*, 27–30.

Palsson, G., Szerszynski, B., Sorlin, S., Marks, J., Avril, B., Crumley, C., . . . Rifka, W. (2013). Reconceptualizing the "Anthropos" in the Anthropocene: Integrating the social sciences and humanities in global environmental change research. *Environmental Science and Policy, 28*, 3–13.

Palter, J. B. (2015). The role of the Gulf Stream in European climate. *Annual Review of Marine Science, 7*, 113–137.

Pan, Y. D., Birdsey, R. A., Fang, J. Y., Houghton, R., Kauppi, P. E., Kurz, W. A., . . . Hayes, D. (2011). A large and persistent carbon sink in the world's forests. *Science, 333*, 988–993.

Pandit, M., and Paudel, K. P. (2016). Water pollution and income relationships: A seemingly unrelated partially linear analysis. *Water Resources Research, 52,* 7668–7689.

Parmesan, C. (2006). Ecological and evolutionary responses to recent climate change. *Annual Review of Ecology Evolution and Systematics, 37,* 637–669.

Parson, R. E. (2008). Useful global-change scenarios: Current issues and challenges. *Environmental Research Letters, 3,* 045016.

Pattberg, P. H. (2007). *Private institutions and global governance.* Northampton, MA: Edward Elgar.

Paul, F. (2014). Global glacial changes. Retrieved from http://www.earsel.org/SIG/Snow -Ice/files/oral_ws2014/Paul_2014_EARSeL.pdf

Pauly, D., Watson, R., and Alder, J. (2005). Global trends in world fisheries: Impacts on marine ecosystems and food security. *Philosophical Transactions of the Royal Society B, 360,* 5–12.

Pavlov, A. A., Kasting, J. F., Brown, L. L., Rages, K. A., and Freedman, R. (2000). Greenhouse warming by CH_4 in the atmosphere of early Earth. *Journal of Geophysical Research–Planets, 105,* 11981–11990.

Payne, J. L., Bush, A. M., Heim, N. A., Knope, M. L., and McCauley, D. J. (2016). Ecological selectivity of the emerging mass extinction in the oceans. *Science, 353,* 1284–1286.

Pearce, F. (2006). *When the rivers run dry.* Boston: Beacon Press.

Peng, P. T., Kumar, A., van den Dool, H., and Barnston, A. G. (2002). An analysis of multimodel ensemble predictions for seasonal climate anomalies. *Journal of Geophysical Research–Atmospheres, 107.* doi: 10.1029/2002jd002712

Peters, W., Jacobson, A. R., Sweeney, C., Andrews, A. E., Conway, T. J., Masarie, K., . . . Tans, P. P. (2007). An atmospheric perspective on North American carbon dioxide exchange: CarbonTracker. *Proceedings of the National Academy of Sciences U.S.A., 104,* 18925–18930.

Petersen, S. V., Dutton, A., and Lohmann, K. C. (2016). End-Cretaceous extinction in Antarctica linked to both Deccan volcanism and meteorite impact via climate change. *Nature Communications, 7.* doi: 10.1038/ncomms12079

Petit, J. R., Jouzel, J., Raynaud, D., Barkov, N. I., Barnola, J. M., Basile, I., . . . Stievenard, M. (1999). Climate and atmospheric history of the past 420,000 years from the Vostok ice core, Antarctica. *Nature, 399,* 429–436.

Pianka, E. R. (1970). On r-selection and K-selection. *American Naturalist, 104,* 592–594.

Pielke, R. A., Pitman, A., Niyogi, D., Mahmood, R., McAlpine, C., Hossain, F., . . . de Noblet, N. (2011). Land use/land cover changes and climate: Modeling analysis and observational evidence. *Wiley Interdisciplinary Reviews–Climate Change, 2,* 828–850.

Pigliucci, M. (2007). Do we need an extended evolutionary synthesis? *Evolution, 61,* 2743–2749.

Polanyi, K. (1944). *The great transformation.* New York: Farrar and Rinehart.

Pound, M. J., Haywood, A. M., Salzmann, U., and Riding, J. B. (2012), Global vegetation dynamics and latitudinal temperature gradients during the mid to late Miocene (15.97–5.33 Ma). *Earth-Science Reviews, 112,* 1–22.

Prigogine, I., and Stengers, I. (1984). *Order out of chaos: Man's new dialogue with nature.* New York: Flamingo.

Prinn, R. G. (2013). Development and application of earth system models. *Proceedings of the National Academy of Sciences U.S.A.*, *110*, 3673–3680.

Provencher-Nolet, L., Bernier, M., and Levesque, E. (2014). Quantification of recent changes to the forest-tundra ecotone through numerical analysis of aerial photographs. *Ecoscience*, *21*, 419–433.

Rahmstorf, S. (2003). The current climate. *Nature*, *421*, 699–699.

Rahmstorf, S., Box, J. E., Feulner, G., Mann, M. E., Robinson, A., Rutherford, S., and Schaffernicht, E. J. (2015). Exceptional twentieth-century slowdown in Atlantic Ocean overturning circulation. *Nature Climate Change*, *5*, 475–480.

Ramankutty, N., Evan, A. T., Monfreda, C., and Foley, J. A. (2008). Farming the planet: Geographic distribution of global agricultural lands in the year 2000. *Global Biogeochemical Cycles*, *22*. doi: 10.1029/2007gb002952

Ramankutty, N., Graumlich, L., Achard, F., Alves, D., Chhabra, A., DeFries, R., . . . Turner, B. L. II. (2006). Global land cover change: Recent progress, remaining challenges. In E. F. Lambin and H. Geist (Eds.), *Land use and land cover change: Local processes, global impacts* (pp. 9–39). New York: Springer Verlag.

Rammer, W., and Seidl, R. (2015). Coupling human and natural systems: Simulating adaptive management agents in dynamically changing forest landscapes. *Global Environmental Change: Human and Policy Dimensions*, *35*, 475–485.

Rampino, M. R., and Caldeira, K. (1994). The Goldilock's problem—climatic evolution and long-term habitability of terrestrial planets. *Annual Review of Astronomy and Astrophysics*, *32*, 83–114.

Randerson, J. T., Lindsay, K., Munoz, E., Fu, W., Moore, J. K., Hoffman, F. M., . . . Doney, S. (2015). Multicentury changes in ocean and land contributions to the climate-carbon feedback. *Global Biogeochemical Cycles*, *29*, 744–759.

Rapacciuolo, G., Maher, S. P., Schneider, A. C., Hammond, T. T., Jabis, M. D., Walsh, R. E., . . . Beissinger, S. R. (2014). Beyond a warming fingerprint: Individualistic biogeographic responses to heterogeneous climate change in California. *Global Change Biology*, *20*, 2841–2855.

Rapport, D. J., and Whitford, W. J. (1999). How ecosystems respond to stress: Common properties of arid and aquatic systems. *Bioscience*, *49*, 193–203.

Raskin, P. (2016). *Journey to Earthland.* Tellus Institute. Retrieved from http://www.greattransition.org/publication/journey-to-earthland

Rauch, J. N., and Pacyna, J. M. (2009). Earth's global Ag, Al, Cr, Cu, Fe, Ni, Pb, and Zn cycles. *Global Biogeochemical Cycles*, *23*. doi: 10.1029/2008gb003376

Raudsepp-Hearne, C., Peterson, G. D., and Bennett, E. M. (2010a). Ecosystem service bundles for analyzing tradeoffs in diverse landscapes. *Proceedings of the National Academy of Sciences U.S.A.*, *107*, 5242–5247.

Raudsepp-Hearne, C., Peterson, G. D., Tengo, M., Bennett, E. M., Holland, T., Benessaiah, K., . . . Pfeifer, L. R. (2010b). Untangling the environmentalist's paradox: Why is human well-being increasing as ecosystem services degrade? *Bioscience, 60,* 576–589.

Raupach, M. R., and Canadell, J. G. (2010). Carbon and the Anthropocene. *Current Opinion in Environmental Sustainability, 2,* 210–218.

Raven, J. A., and Falkowski, P. G. (1999). Oceanic sinks for atmospheric CO_2. *Plant Cell and Environment, 22,* 741–755.

Raymo, M. E., and Ruddiman, W. F. (1992). Tectonic forcing of late Cenozoic climate. *Nature, 359,* 117–122.

Reed, M. S. (2008). Stakeholder participation for environmental management: A literature review. *Biological Conservation, 141,* 2417–2431.

Reichstein, M., Ciais, P., Papale, D., Valentini, R., Running, S., Viovy, N., . . . Pilegaard, K. (2006). Reduction of ecosystem productivity and respiration during the European summer 2003 climate anomaly: A joint flux tower, remote sensing and modelling analysis. *Global Change Biology, 12,* 1–18.

Retallack, G. J. (2002). Carbon dioxide and climate over the past 300 Myr. *Philosophical Transactions of the Royal Society of London Series A–Mathematical Physical and Engineering Sciences, 360,* 659–673.

Retallack, G. J. (2009). Greenhouse crises of the past 300 million years. *Geological Society of America Bulletin, 121,* 1441–1455.

Retallack, G. J., and Jahren, A. H. (2008). Methane release from igneous intrusion of coal during late Permian extinction events. *Journal of Geology, 116,* 1–20.

Revelle, R., and Suess, H. E. (1957). Carbon dioxide exchange between atmosphere and ocean and the question of an increase of atmospheric CO_2 during the past decades. *Tellus, 9,* 18–27.

Riahi, K., Rao, S., Krey, V., Cho, C. H., Chirkov, V., Fischer, G., . . . Rafaj, P. (2011). RCP 8.5-A scenario of comparatively high greenhouse gas emissions. *Climatic Change, 109,* 33–57.

Ripple, W. J., and Beschta, R. L. (2004). Wolves and the ecology of fear: Can predation risk structure ecosystems? *BioScience, 54,* 755–766.

Ripple, W. J., and Beschta, R. L. (2005). Linking wolves and plants: Aldo Leopold on trophic cascades. *BioScience, 55,* 613–621.

Ripple, W. J., Chapron, G., Lopez-Bao, J. V., Durant, S. M., Macdonald, D. W., Lindsey, P. A., . . . Zhang, L. (2016). Saving the world's terrestrial megafauna. *Bioscience, 66,* 807–812.

Rittel, H. W. J., and Webber, M. M. (1973). Dilemmas in a general theory of planning. *Policy Sciences, 4,* 155–169.

Robertson, R. (1992). *Globalization: Social theory and global culture.* Thousand Oaks, CA: Sage.

Robinson, A., Calov, R., and Ganopolski, A. (2012). Multistability and critical thresholds of the Greenland ice sheet. *Nature Climate Change, 2,* 429–432.

Robinson, J. M. (1990). Lignin, land plants, and fungi—biological evolution affecting Phanerozoic oxygen balance. *Geology, 18,* 607–610.

Robock, A. (2000). Volcanic eruptions and climate. *Reviews of Geophysics, 38*, 191–219.

Rockström, J., Steffen, W., Noone, K., Persson, A., Chapin, F. S., Lambin, E. F., . . . Foley, J. A. (2009). A safe operating space for humanity. *Nature, 461*, 472–475.

Romme, W. H., Allen, C. D., Bailey, J. D., Baker, W. L., Bestelmeyer, B. T., Brown, P. M., . . . Weisberg, P. J. (2007). Historical and modern disturbance regimes, stand structures, and landscape dynamics in pinon-juniper vegetation of the western United States. *Rangeland Ecology and Management, 62*, 203–222.

Rounsevell, M. D. A., Robinson, D. T., and Murray-Rust, D. (2012). From actors to agents in socio-ecological systems models. *Philosophical Transactions of the Royal Society B–Biological Sciences, 367*, 259–269.

Rowland, F. S. (2006). Stratospheric ozone depletion. *Philosophical Transactions of the Royal Society B–Biological Sciences, 361*, 769–790.

Roxy, M. K., Modi, A., Murtugudde, R., Valsala, V., Panickal, S., Kumar, S. P., . . . Lévy, M. (2016). A reduction in marine primary productivity driven by rapid warming over the tropical Indian Ocean. *Geophysical Research Letters, 43*, 826–833.

Roy, J., Saugier, B., and Mooney, H. A. (2001). *Terrestrial global productivity.* Cambridge, MA: Academic Press.

Royer, D. L., Berner, R. A., Montanez, I. P., Tabor, N. J., and Beerling, D. J. (2004). CO_2 as a primary driver of Phanerozoic climate. *GSA Today, 14*, 4–10.

Ruddiman, W. F. (Ed.). (1997). *Tectonic uplift and climate change.* New York: Plenum.

Ruddiman, W. F. (2005). *Plows, plagues, and petroleum: How humans took control of climate.* Princeton, NJ: Princeton University Press.

Ruddiman, W. F., Fuller, D. Q., Kutzbach, J. E., Tzedakis, P. C., Kaplan, J. O., Ellis, E. C., . . . Woodbridge, J. (2016). Late Holocene climate: Natural or anthropogenic? *Reviews of Geophysics, 54*, 93–118.

Rudel, T. K., Coomes, O. T., Moran, E., Achard, F., Angelsen, A., Xu, J. C., and Lambin, E. (2005). Forest transitions: Towards a global understanding of land use change. *Global Environmental Change: Human and Policy Dimensions, 15*, 23–31.

Rudel, T. K., Defries, R., Asner, G. P., and Laurance, W. F. (2009). Changing drivers of deforestation and new opportunities for conservation. *Conservation Biology, 23*, 1396–1405.

Running, S. W. (2014). A regional look at HANPP: Human consumption is increasing, NPP is not. *Environmental Research Letters, 9.* doi: 10.1088/1748–9326/9/11/111003

Running, S. W., Nemani, R. R., Heinsch, F. A., Zhao, M., Reeves, M., and Hashimoto, H. (2004). A continuous satellite-derived measure of global terrestrial production. *BioScience, 54*, 547–560.

Saatchi, S. S., Harris, N. L., Brown, S., Lefsky, M., Mitchard, E. T. A., Salas, W., . . . Morel, A. (2011). Benchmark map of forest carbon stocks in tropical regions across three continents. *Proceedings of the National Academy of Sciences U.S.A., 108*, 9899–9904.

Sachs, J. D. (2015). *The age of sustainable development.* New York: Columbia University Press.

Sage, R. F., and Monson, R. K. (1998). *C4 plant biology*. Cambridge, MA: Academic Press.

Saha, S., Moorthi, S., Pan, H. L., Wu, X. R., Wang, J. D., Nadiga, S., . . . Goldberg, M. (2010). The NCEP climate forecast system reanalysis. *Bulletin of the American Meteorological Society, 91*, 1015–1057.

Salzmann, U., Haywood, A. M., and Lunt, D. J. (2009) The past is a guide to the future? Comparing Middle Pliocene vegetation with predicted biome distributions for the twenty-first century. *Philosophical Transactions of the Royal Society A–Mathematical Physical and Engineering Sciences, 367*, 189–204.

Samir, K. C., and Lutz, W. G. (2014). Demographic scenarios by age, sex and education corresponding to the SSP narratives. *Population and Environment, 35*, 243–260.

Sampson, P. R., and Pitt, D. (Eds.). (1999). *The biosphere and noösphere reader*. Abingdon-on-Thames, UK: Routledge.

Saugier, B., Roy, J., and Mooney, H. A. (2001). Estimations of global terrestrial productivity: Converging toward a single number. In J. Roy, B. Saugier, and H. A. Mooney (Eds.), *Terrestrial global productivity* (pp. 543–557). Cambridge, MA: Academic Press.

Saunders, H. (2016). Does capitalism require endless growth? *Breakthrough Journal*. Retrieved from https://thebreakthrough.org/index.php/journal/issue-6/does-capitalism-require-endless-growth

Schaller, M. F., Wright, J. D., and Kent, D. V. (2015). A 30 Myr record of Late Triassic atmospheric pCO(2) variation reflects a fundamental control of the carbon cycle by changes in continental weathering. *Geological Society of America Bulletin, 127*, 661–671.

Schellnhuber, H. J. (1999). Earth system analysis and the second Copernican Revolution. *Nature, 402*, C19-C23.

Schellnhuber, H. J., Crutzen, P. J., Clark, W. C., Claussen, M., and Held, H. (Eds.). (2004). *Earth system analysis for sustainability*. Cambridge, MA: MIT Press.

Schiermeier, Q. (2015). Why the Pope's letter on climate change matters. *Nature*. doi:10.1038/nature.2015.17800

Schimel, D. S., House, J. I., Hibbard, K. A., Bousquet, P., Ciais, P., Peylin, P., . . . Wirth, C. (2001). Recent patterns and mechanisms of carbon exchange by terrestrial ecosystems. *Nature, 414*, 169–172.

Schimel, D., Pavlick, R., Fisher, J. B., Asner, G. P., Saatchi, S., Townsend, P., . . . Cox, P. (2015). Observing terrestrial ecosystems and the carbon cycle from space. *Global Change Biology, 21*, 1762–1776.

Schlesinger, W. H. (2013). Requiem for a grand theory. *Nature Climate Change, 3*, 697–697.

Schlesinger, W. H., and Bernhart, E. S. (2013). *Biogeochemistry: An analysis of global change*. Cambridge, MA: Academic Press.

Schleussner, C. F., Lissner, T. K., Fischer, E. M., Wohland, J., Perrette, M., Golly, A., . . . Schaeffer, M. (2016). Differential climate impacts for policy-relevant limits to global warming: The case of 1.5 °C and 2 °C. *Earth System Dynamics, 7*, 327–351.

Schlosberg, D., and Rinfret, S. (2008). Ecological modernisation, American style. *Environmental Politics, 17*, 254–275.

Schmittner, A., Urban, N. M., Shakun, J. D., Mahowald, N. M., Clark, P. U., Bartlein, P. J., . . . Rosell-Mele, A. (2011). Climate sensitivity estimated from temperature reconstructions of the Last Glacial Maximum. *Science, 334*, 1385–1388.

Schneider, S. H., and Boston, P. J. (Eds.). (1991). *Scientists on Gaia.* Cambridge, MA: MIT Press.

Schneider, S. H., Miller, J. R., Crist, E., and Boston, P. J. (Eds.). (2004). *Scientists debate Gaia.* Cambridge, MA: MIT Press.

Schobben, M., Stebbins, A., Ghaderi, A., Strauss, H., Korn, D., and Korte, C. (2015). Flourishing ocean drives the end-Permian marine mass extinction. *Proceedings of the National Academy of Sciences U.S.A., 112*, 10298–10303.

Scholes, R. J. (2016). Climate change and ecosystem services. *Wiley Interdisciplinary Reviews–Climate Change, 7*, 537–550.

Scholes, R. J., Walters, M., Turak, E., Saarenmaa, H., Heip, C. H. R., Tuama, E. O., . . . Geller, G. (2012). Building a global observing system for biodiversity. *Current Opinion in Environmental Sustainability, 4*, 139–146.

Schug, T. T., Johnson, A. F., Birnbaum, L. S., Colborn, T., Guillette, L. J., Crews, D. P., . . . Heidel, J. J. (2016). Microreview. Endocrine disruptors: Past lessons and future directions. *Molecular Endocrinology, 30*, 833–847.

Schuur, E. A. G., McGuire, A. D., Schadel, C., Grosse, G., Harden, J. W., Hayes, D. J., . . . Vonk, J. E. (2015). Climate change and the permafrost carbon feedback. *Nature, 520*, 171–179.

Schwartzman, D. (1999). *Life, temperature, and the Earth.* New York: Columbia University Press.

Scripps Oceanography. (2016). Global environmental monitoring. Retrieved from https://scripps.ucsd.edu/research/themes/global-environmental-monitoring

Seager, R., Ting, M., Kushnir, Y., Lu, J., Vecchi, G., Huang, H., . . . Naik, N. (2007). Model projections of an imminent transition to a more arid climate in southwestern North America. *Science, 316*, 1181–1184.

SEDAC. (2016). NASA Socioeconomic Data and Applications Center. Retrieved from http://sedac.ciesin.org/entri/guides/sec3-T+E.html

Seddon, N., Mace, G. M., Naeem, S., Tobias, J. A., Pigot, A. L., Cavanagh, R., . . . Walpole, M. (2016). Biodiversity in the Anthropocene: Prospects and policy. *Proceedings of the Royal Society B–Biological Sciences, 283*, 20162094.

Sedgh, G., Singh, S., and Hussain, R. (2014). Intended and unintended pregnancies worldwide in 2012 and recent trends. *Studies in Family Planning, 45*, 301–314.

Serra-Diaz, J. M., Franklin, J., Dillon, W. W., Syphard, A. D., Davis, F. W., and Meentemeyer, R. K. (2016). California forests show early indications of both range shifts and local persistence under climate change. *Global Ecology and Biogeography, 25*, 164–175.

Seyfang, G. (2003). Environmental mega-conferences—from Stockholm to Johannesburg and beyond. *Global Environmental Change: Human and Policy Dimensions, 13,* 223–228.

Shaffer, G., Huber, M., Rondanelli, R., and Pedersen, J. O. P. (2016). Deep time evidence for climate sensitivity increase with warming. *Geophysical Research Letters, 43,* 6538–6545.

Sherwood, S. C., and Huber, M. (2010) An adaptability limit to climate change due to heat stress. *Proceedings of the National Academy of Sciences U.S.A., 107,* 9552–9555.

Shevliakova, E., Stouffer, R. J., Malyshev, S., Krasting, J. P., Hurtt, G. C., and Pacala, S. W. (2013). Historical warming reduced due to enhanced land carbon uptake. *Proceedings of the National Academy of Sciences U.S.A., 110,* 16730–16735.

Shindell, D., Kuylenstierna, J. C. I., Vignati, E., van Dingenen, R., Amann, M., Klimont, Z., . . . Fowler, D. (2012). Simultaneously mitigating near-term climate change and improving human health and food security. *Science, 335,* 183–189.

Siegel, D. A., Buesseler, K. O., Doney, S. C., Sailley, S. F., Behrenfeld, M. J., and Boyd, P. W. (2014). Global assessment of ocean carbon export by combining satellite observations and food-web models. *Global Biogeochemical Cycles, 28,* 181–196.

Six, K. D., Kloster, S., Ilyina, T., Archer, S. D., Zhang, K., and Maier-Reimer, E. (2013). Global warming amplified by reduced sulphur fluxes as a result of ocean acidification. *Nature Climate Change, 3,* 975–978.

SLRMap. (2017). Global sea level rise map. Retrieved from http://geology.com/sea-level-rise/

SMAP. (2017). Surface Moisture Active Passive. Retrieved from https://smap.jpl.nasa.gov/

Smil, V. (2002). *The Earth's biosphere.* Cambridge, MA: MIT Press.

Smith, C. L., Lopes, V. L., and Carrejo, F. M. (2011). Recasting paradigm shift: "true" sustainability and complex systems. *Human Ecology Review, 18,* 67–74.

Smith, P., Davis, S. J., Creutzig, F., Fuss, S., Minx, J., Gabrielle, B., . . . Cho, Y. (2016). Biophysical and economic limits to negative CO_2 emissions. *Nature Climate Change, 6,* 42–50.

Smith, R. (2015). *Green capitalism: The god that failed.* Norcross, GA: College Publications.

Smith, W. K., Reed, S. C., Cleveland, C. C., Ballantyne, A. P., Anderegg, W. R. L., Wieder, W. R., . . . Running, S. W. (2016). Large divergence of satellite and Earth system model estimates of global terrestrial CO_2 fertilization. *Nature Climate Change, 6,* 306–310.

Solomon, S., Ivy, D. J., Kinnison, D., Mills, M. J., Neely, R. R., and Schmidt, A. (2016). Emergence of healing in the Antarctic ozone layer. *Science, 353,* 269–274.

SOTC. (2016). State of the climate report. Retrieved from https://www.ncdc.noaa.gov/sotc/global/201513

Spaargaen, G., and Mol, A. P. J. (2008). Greening global consumption: Redefining politics and authority. *Global Environmental Change: Human and Policy Dimensions, 18,* 350–359.

Spaargaen, G., Mol, A. P. J., and Buttel, F. H. (Eds.). (2000). *Environment and global modernity.* Thousand Oaks, CA: Sage.

Steadman, W. D. (2006). *Extinction and biogeography of tropical Pacific birds.* Chicago: University of Chicago Press.

Steffen, W., Broadgate, W., Deutsch, L., Gaffney, O., and Ludwig, C. (2015a). The trajectory of the Anthropocene: The Great Acceleration. *Anthropocene Review, 2,* 81–98.

Steffen, W., Richardson, K., Rockström, J., Cornell, S. E., Fetzer, I., Bennett, E. M., . . . Sörlin, S. (2015b). Sustainability. Planetary boundaries: Guiding human development on a changing planet. *Science, 347.* doi: 10.1126/science.1259855

Steinacher, M., Joos, F., Frolicher, T. L., Bopp, L., Cadule, P., Cocco, V., . . . Segschneider, J. (2010). Projected 21st century decrease in marine productivity: A multi-model analysis. *Biogeosciences, 7,* 979–1005.

Steinfeld, H., Gerber, P., Wassenaar, T., Castel, V., Rosales, M., and de Haan, C. (2006). *Livestock's long shadow: Environmental issues and options.* Rome, Italy: FAO.

Strassberg, G., Scanlon, B. R., and Chambers, D. (2009). Evaluation of groundwater storage monitoring with the GRACE satellite: Case study of the High Plains aquifer, central United States. *Water Resources Research, 45.* doi: 10.1029/2008wr006892

Summerhayes, C. P. (2015). *Earth's climate evolution.* Hoboken, NJ: Wiley Blackwell.

SURFRAD. (2014). Surface Radiation Network. Retrieved from http://www.esrl.noaa.gov/gmd/grad/surfrad/

Suri, V., and Chapman, D. (1998). Economic growth, trade and energy: Implications for the environmental Kuznets curve. *Ecological Economics, 25,* 195–208.

Sustainable Development. (2017). United Nations sustainable development knowledge platform. Retrieved from https://sustainabledevelopment.un.org/milestones/unced.

Swenson, W., Wilson, D. S., and Elias, R. (2000). Artificial ecosystem selection. *Proceedings of the National Academy of Science U.S.A., 97,* 9110–9114.

Tainter, J. A. (1988). *The collapse of complex societies.* Cambridge: Cambridge University Press.

Tainter, J. A. (2011). Energy, complexity, and sustainability: A historical perspective. *Environmental Innovation and Societal Transitions, 1,* 89–95.

Taylor, K. E., Stouffer, R. J., and Meehl, G. A. (2012). An overview of CMIP5 and the experiment design. *Bulletin American Meteorological Society, 93,* 485–498.

Teilhard de Chardin, P. (1959/1955). *The phenomenon of man.* New York: Collins.

Thomas, J. W., Franklin, J. F., Gordon, J., and Johnson, K. N. (2006). The Northwest Forest Plan: Origins, components, implementation experience, and suggestions for change. *Conservation Biology, 20,* 277–287.

Thomson, A. M., Calvin, K. V., Smith, S. J., Kyle, G. P., Volke, A., Patel, P., . . . Edmonds, J. A. (2011). RCP4.5: A pathway for stabilization of radiative forcing by 2100. *Climatic Change, 109,* 77–94.

Thornton, P. K. (2010). Livestock production: recent trends, future prospects. *Philosophical Transactions of the Royal Society B–Biological Sciences, 365,* 2853–2867.

Tingey, D. T., Turner, D. P., and Weber, J. A. (1991). Factors controlling the emissions of monoterpenes and other volatile organic compounds. In T. D. Sharkey, E. A. Holland, and H. A. Mooney (Eds.), *Trace gas emissions by plants* (pp. 93–120). Cambridge, MA: Academic Press.

Tng, D. Y. P., Murphy, B. P., Weber, E., Sanders, G., Williamson, G. J., Kemp, J., and Bowman, D. (2012). Humid tropical rain forest has expanded into eucalypt forest and savanna over the last 50 years. *Ecology and Evolution, 2*, 34–45.

Tokarska, K. B., Gillett, N. P., Weaver, A. J., Arora, V. K., and Eby, M. (2016). The climate response to five trillion tonnes of carbon. *Nature Climate Change, 6*, 851–856.

Tollefson, J. (2015). Battle for the Amazon. *Nature, 520*, 20–23.

Townshend, J. R. C., and Justice, C. O. (1988). Selecting the spatial resolution of satellite sensors required for global monitoring land transformations. *International Journal of Remote Sensing, 9*, 187–236.

Trenberth, K. E., Zhang, Y., Fasullo, J. T., and Taguchi, S. (2015). Climate variability and relationships between top-of-atmosphere radiation and temperatures on Earth. *Journal of Geophysical Research–Atmospheres, 120*, 3642–3659.

Turner, A. J., Jacob, D. J., Benmergui, J., Wofsy, S. C., Maasakkers, J. D., Butz, A., . . . Biraud, S. C. (2016). A large increase in US methane emissions over the past decade inferred from satellite data and surface observations. *Geophysical Research Letters, 43*, 2218–2224.

Turner, B. L. (2010). Vulnerability and resilience: Coalescing or paralleling approaches for sustainability science? *Global Environmental Change: Human and Policy Dimensions, 20*, 570–576.

Turner, B. L. II, Clark, W. C., Kates, R. W., and Richards, J. F. (1993). *The Earth as transformed by human action: Global and regional changes in the biosphere over the past 300 Years*. Cambridge: Cambridge University Press.

Turner, D. P. (2005). Thinking at the global scale. *Global Ecology and Biogeography, 14*, 505–508.

Turner, D. P. (2011). Global vegetation monitoring: Towards a sustainable technobiosphere. *Frontiers in Ecology and the Environment, 9*, 111–116.

Turner, D. P., Baglio, J. V., Wones, A. G., Pross, D., Vong, R., McVeety, B. D., and Phillips, D. L. (1991). Climate change and isoprene emissions from vegetation. *Chemosphere, 23*, 37–56.

Turner, D. P., Conklin, D. R., and Bolte, J. P. (2015). Projected climate change impacts on forest land cover and land use over the Willamette River Basin, Oregon, USA. *Climatic Change, 133*, 335–348.

Turner, D, P., Conklin, D. R., Vache, K. B., Schwartz, C., Nolin, A. W., Chang H, . . . Bolte J. P. (2016). Assessing mechanisms of climate change impact on the upland forest water balance of the Willamette River Basin, Oregon. *Ecohydrology, 10*. doi: 10.1002/eco.1776

Turner, D. P., Ollinger, S. V., and Kimball, J. S. (2004). Integrating remote sensing and ecosystem process models for landscape to regional scale analysis of the carbon cycle. *BioScience, 54*, 573–584.

Turner, D. P., Ritts, W. D., Cohen, W. B., Gower, S. T., Running, S. W., Zhao, M. S., . . . Ahl, D. E. (2006). Evaluation of MODIS NPP and GPP products across multiple biomes. *Remote Sensing of Environment, 102*, 282–292.

Turner, D. P., Ritts, W. D., Cohen, W. B., Maeirsperger, T., Gower, S. T., Kirschbaum, A., . . . Gamon, J. A. (2005). Site-level evaluation of satellite-based global terrestrial gross primary production and net primary production monitoring. *Global Change Biology, 11*, 666–684.

Turner, D. P., Urbanski, S., Bremer, D., Wofsy, S. C., Meyers, T., Gower, S. T., and Gregory, M. (2003). A cross-biome comparison of daily light use efficiency for gross primary production. *Global Change Biology, 9*, 383–395.

Turner, M. G., Gardner, R. H., and O'Neill, R. V. (2003). *Landscape ecology in theory and practice: Pattern and process.* New York: Springer.

UNEP. (2011). Towards a green economy: Pathways to sustainable development and poverty eradication—A synthesis for policy makers. Retrieved from http://www.unep.org/greeneconomy

UNEP. (2012). State of the science of endocrine disrupting chemicals 2012: Summary for decision-makers. Retrieved from http://www.who.int/ceh/publications/endocrine/en/

UNEP. (2016). United Nations Environmental Programme. World conservation monitoring Centre. Retrieved from https://www.unep-wcmc.org/featured-projects/mapping-the-worlds-special-places

United Nations. (1992). Rio Declaration on Environment and Development. Retrieved from http://www.un.org/documents/ga/conf151/aconf15126-1annex1.htm

United Nations. (2005). World urbanization prospects: The 2005 revision. Retrieved from http://www.un.org/esa/population/publications/WUP2005/2005WUPHighlights_Final_Report.pdf

USCB. (2017). U.S. Census Bureau. International programs. Retrieved from http://www.census.gov/population/international/data/worldpop/graph_growthrate.php

USDA. (2012). Forest plan revision collaborative guide. Retrieved from http://www.csus.edu/ccp/presentations/stelprdb5422900.pdf

USDA. (2016a). Organizing for public engagement under the 2012 planning rule lessons learned. Retrieved from https://www.fs.fed.us/emc/nfma/collaborative_processes/documents/OrgforPubEngmntunderthe2102PlngRleLssnsLrnd.pdf

USDA. (2016b). U.S. Department of Agriculture Resources Planning Act assessment. Retrieved from http://www.fs.fed.us/research/rpa/

USGBC. (2017). United States Green Building Council. Retrieved from http://www.usgbc.org/

van der Werf, G. R., Dempewolf, J., Trigg, S. N., Randerson, J. T., Kasibhatla, P. S., Gigliof, L., . . . DeFries, R. S. (2008). Climate regulation of fire emissions and deforestation in equatorial Asia. *Proceedings of the National Academy of Sciences U.S.A., 105*, 20350–20355.

van Lierop, P., Lindquist, E., Sathyapala, S., and Franceschini, G. (2015). Global forest area disturbance from fire, insect pests, diseases and severe weather events. *Forest Ecology and Management, 352*, 78–88.

van Mantgem, P. J., Stephenson, N. L., Byrne, J. C., Daniels, L. D., Franklin, J. F., Fule, P. Z., . . . Veblen, T. T. (2009). Widespread increase of tree mortality rates in the western United States. *Science, 323*, 521–524.

van Vuuren, D. P., Edmonds, J., Kainuma, M., Riahi, K., Thomson, A., Hibbard, K., . . . Rose, S. (2011). The representative concentration pathways: an overview. *Climatic Change, 109*, 5–31.

van Vuuren, D. P., Kok, M., Lucas, P. L., Prins, A. G., Alkemade, R., van den Berg, M., . . . Stehfest, E. (2015). Pathways to achieve a set of ambitious global sustainability objectives by 2050: Explorations using the IMAGE integrated assessment model. *Technological Forecasting and Social Change, 98*, 303–323.

Vandenburg, W. H. (1985). *The growth of minds and cultures.* Toronto: University of Toronto Press.

Vellinga, M., and Wood, R. A. (2008). Impacts of thermohaline circulation shutdown in the twenty-first century. *Climatic Change, 91*, 43–63.

Vernadsky, V. I. (1945). The biosphere and the noösphere. *American Scientist, 33*, 1–12.

Visser, M. E., and Both, C. (2005). Shifts in phenology due to global climate change: The need for a yardstick. *Proceedings of the Royal Society B–Biological Sciences, 272*, 2561–2569.

Vitousek, P. M., Aber, J. D., Howarth, R. W., Likens, G. E., Matson, P. A., Schindler, D. W., . . . Tilman, D. (1997a). Human alteration of the global nitrogen cycle: Sources and consequences. *Ecological Applications, 7*, 737–750.

Vitousek, P. M., Mooney, H. A., Lubchenco, J., and Melillo, J. M. (1997b). Human domination of Earth's ecosystem. *Science, 277*, 494–499.

Volk, T. (1995). *Metapatterns.* New York: Columbia University Press.

Volk, T. (1998). *Gaia's body: Towards a physiology of Earth.* New York: Springer-Verlag.

Wackernagel, M., and Rees, W. E. (1997). Perceptual and structural barriers to investing in natural capital: Economics from an ecological footprint perspective. *Ecological Economics, 20*, 3–24.

Walker, B., Holling, C. S., Carpenter, S. R., and Kinzig, A. (2004). Resilience, adaptability and transformability in social-ecological systems. *Ecology and Society, 9*(5).

Walker, J. C. G. (1985). Carbon dioxide on the early Earth. *Origins of Life, 16*, 117–127.

Walker, J. C. G., Hays, P. B., and Kasting, J. F. (1981). A negative feedback mechanism for the long-term stabilization of Earth's surface temperature. *Journal of Geophysical Research, 86*, 9776–9782.

Wang, S. Y., Hipps, L., Gillies, R. R., and Yoon, J. H. (2014). Probable causes of the abnormal ridge accompanying the 2013-2014 California drought: ENSO precursor and anthropogenic warming footprint. *Geophysical Research Letters, 41*, 3220–3226.

Wang, W. L., Ciais, P., Nemani, R. R., Canadell, J. G., Piao, S. L., Sitch, S., . . . (2013). Variations in atmospheric CO2 growth rates coupled with tropical temperature. *Proceedings of the National Academy of Sciences U.S.A., 110*, 13061–13066.

Ward, P. (2009). *The Medea hypothesis.* Princeton, NJ: Princeton University Press.

Ward, P., Labandeira, C., Laurin, M., and Berner, R. A. (2006). Confirmation of Romer's gap as a low oxygen interval constraining the timing of initial arthropod and vertebrate terrestrialization. *Nature, 103,* 16818–16822.

Watson, P. J. (2016). A new perspective on global mean sea level (GMSL) acceleration. *Geophysical Research Letters, 43,* 6478–6484.

Weaver, A. J., Sedlacek, J., Eby, M., Alexander, K., Crespin, E., Fichefet, T., . . . Tokos, M. (2012). Stability of the Atlantic meridional overturning circulation: A model intercomparison. *Geophysical Research Letters, 39.* doi: 10.1029/2012gl053763

Westerling, A. L., Hidalgo, H. G., Cayan, D. R., and Swetnam, T. W. (2006). Warming and earlier spring increase western US forest wildfire activity. *Science, 313,* 940–943.

Westley, F. R., Tjornbo, O., Schultz, L., Olsson, P., Folke, C., Crona, B., and Bodin, O. (2013). A theory of transformative agency in linked social-ecological systems. *Ecology and Society, 18.* doi: 10.5751/es-05072–180327

Wigley, T. M. L., Ramaswamy, V., Christy, J. R., Lanzante, J. R., Mears, C. A., Santer, B. D., and Folland, C. K. (2006). Executive summary: Temperature trends in the lower atmosphere—Understanding and reconciling differences. United States Global Climate Change Research Program. Retrieved from https://www.climatecommunica tion.org/wp-content/uploads/2011/08/temptrends.pdf

Wilcox, L. J., Highwood, E. J., and Dunstone, N. J. (2013). The influence of anthropogenic aerosol on multi-decadal variations of historical global climate. *Environmental Research Letters, 8.* doi: 10.1088/1748–9326/8/2/024033

Williams, H. T. P., and Lenton, T. M. (2008). Environmental regulation in a network of simulated microbial ecosystems. *Proceedings of the National Academy of Sciences U.S.A., 105,* 10432–10437.

Williams, M. (1989). *Americans and their forests.* Cambridge: Cambridge University Press.

Williams, M., Zalasiewicz, J. A., Haff, P. K., Schwagerl, C., Barnosky, A. D., and Ellis, E. C. (2015). The Anthropocene biosphere. *Anthropocene Review, 2,* 196–219.

Wilson, D. S. (2001). Evolutionary biology: Struggling to escape exclusively individual selection. *Quarterly Review of Biology, 76,* 199–205.

Wilson, E. O. (1999). *Consilience.* New York: Vintage.

Wilson, E. O. (2016). *Half-Earth.* New York: Norton.

Wilson, E. O., and Hölldobler, B. (2008). *The superorganism. The beauty, elegance, and strangeness of insect societies.* New York: Norton.

Wilson, J., Yan, L. Y., and Wilson, C. (2007). The precursors of governance in the Maine lobster fishery. *Proceedings of the National Academy of Sciences U.S.A., 104,* 15212–15217.

Wiltshire, A., Gornall, J., Booth, B., Dennis, E., Falloon, P., Kay, G.., and Betts, R. (2013). The importance of population, climate change and CO_2 plant physiological forcing in determining future global water stress. *Global Environmental Change: Human and Policy Dimensions, 23,* 1083–1097.

Wofsy, S. C., Goulden, M. L., Munger, J. W., Fan, S. M., Bakwin, P. S., Daube, B. C., . . . Bazzaz, F. A. (1993). Net exchange of CO_2 in a mid-latitude forest. *Science, 260,* 1314–1317.

Woodcock, C. E., Allen, R., Anderson, M., Belward, A., Bindschadler, R., Cohen, W., . . . Wynne, R. (2008). Free access to Landsat imagery. *Science, 320,* 1011–1011.

World Bank. (2013a). Fish to 2030: Prospects for fisheries and aquaculture. (Agriculture and Environmental Services Discussion Paper No. 3.) Washington, DC: World Bank Group. Retrieved from http://documents.worldbank.org/curated/en/2013/12/18882045/fish-2030-prospects-fisheries-aquaculture

World Bank. (2013b). The World Bank. Retrieved from http://www.worldbank.org/en/topic/environment

WTO. (2017). World Trade Organization. Retrieved from https://www.wto.org/english/thewto_e/whatis_e/inbrief_e/inbr03_e.htm

Wulder, M. A., and Coops, N. C. (2014). Make Earth observations open access. *Nature, 513,* 30–31.

WW2100. (2017). Willamette Water 2100 Project. Retrieved from http://inr.oregonstate.edu/ww2100

Wylie, P. (1971). Cultural evolution—Fatal fallacy. *Bioscience, 21,* 729–732.

Young, O. R. (Ed.). (1997). *Global governance.* Cambridge, MA: MIT Press.

Young, O. R., Berkhout, F., Gallopin, G. C., Janssen, M. A., Ostrom, E., and Leeuw, S. V. D. (2006). The globalization of socio-ecological systems: An agenda for scientific research. *Global Environmental Change: Human and Policy Dimensions, 16,* 304–316.

Yu, M., Wang, G. L., Parr, D., Ahmed, K. F. (2014). Future changes of the terrestrial ecosystem based on a dynamic vegetation model driven with RCP8.5 climate projections from 19 GCMs. *Climatic Change, 127,* 257–271.

Zachos, J. C., Dickens, G. R., and Zeebe, R. E. (2008). An early Cenozoic perspective on greenhouse warming and carbon-cycle dynamics. *Nature, 451,* 279–283.

Zalasiewicz, J., Williams, M., Waters, C. N., Barnosky, A. D., Palmesino, J., Rönnskog, A. S., . . . Wolfe, A. P. (2016). Scale and diversity of the physical technosphere: A geological perspective. *Anthropocene Review, 4,* 9–22. doi: 10.1177/2053019616677743

Zeebe, R. E., Ridgwell, A., and Zachos, J. C. (2016). Anthropogenic carbon release rate unprecedented during the past 66 million years. *Nature Geoscience, 9,* 325–329.

Zhang, M. H., Lin, W. Y., Klein, S. A., Bacmeister, J. T., Bony, S., Cederwall, R. T., . . . Zhang, J. H. (2005). Comparing clouds and their seasonal variations in 10 atmospheric general circulation models with satellite measurements. *Journal of Geophysical Research–Atmospheres, 110.* doi: 10.1029/2004jd005021

Zhao, M., Golaz, J. C., Held, I. M., Ramaswamy, V., Lin, S. J., Ming, Y., . . . Guo, H. (2016). Uncertainty in model climate sensitivity traced to representations of cumulus precipitation microphysics. *Journal of Climate, 29,* 543–560.

Zhao, M. S., and Running, S. W. (2010). Drought-induced reduction in global terrestrial net primary production from 2000 through 2009. *Science, 329,* 940–943.

Zhu, K., Woodall, C. W., and Clark, J. S. (2012). Failure to migrate: Lack of tree range expansion in response to climate change. *Global Change Biology, 18,* 1042–1052.

Zhu, Z. C., Piao, S. L., Myneni, R. B., Huang, M. T., Zeng, Z. Z., Canadell, J. G., . . . Zeng, N. (2016). Greening of the Earth and its drivers. *Nature Climate Change, 6,* 791–795.

INDEX